T0055018

The History of Chemistry: A Very Short Introduction

VERY SHORT INTRODUCTIONS are for anyone wanting a stimulating and accessible way into a new subject. They are written by experts, and have been translated into more than 40 different languages.

The series began in 1995, and now covers a wide variety of topics in every discipline. The VSI library now contains over 450 volumes—a Very Short Introduction to everything from Psychology and Philosophy of Science to American History and Relativity—and continues to grow in every subject area.

Very Short Introductions available now:

ACCOUNTING Christopher Nobes
ADVERTISING Winston Fletcher
AFRICAN AMERICAN RELIGION
 Eddie S. Glaude Jr.
AFRICAN HISTORY John Parker and
 Richard Rathbone
AFRICAN RELIGIONS Jacob K. Olupona
AGNOSTICISM Robin Le Poidevin
ALEXANDER THE GREAT Hugh Bowden
ALGEBRA Peter M. Higgins
AMERICAN HISTORY Paul S. Boyer
AMERICAN IMMIGRATION
 David A. Gerber
AMERICAN LEGAL HISTORY
 G. Edward White
AMERICAN POLITICAL HISTORY
 Donald Critchlow
AMERICAN POLITICAL PARTIES
 AND ELECTIONS L. Sandy Maisel
AMERICAN POLITICS Richard M. Valelly
THE AMERICAN PRESIDENCY
 Charles O. Jones
THE AMERICAN REVOLUTION
 Robert J. Allison
AMERICAN SLAVERY
 Heather Andrea Williams
THE AMERICAN WEST Stephen Aron
AMERICAN WOMEN'S HISTORY
 Susan Ware
ANAESTHESIA Aidan O'Donnell
ANARCHISM Colin Ward
ANCIENT ASSYRIA Karen Radner
ANCIENT EGYPT Ian Shaw
ANCIENT EGYPTIAN ART AND
 ARCHITECTURE Christina Riggs

ANCIENT GREECE Paul Cartledge
THE ANCIENT NEAR EAST
 Amanda H. Podany
ANCIENT PHILOSOPHY Julia Annas
ANCIENT WARFARE Harry Sidebottom
ANGELS David Albert Jones
ANGLICANISM Mark Chapman
THE ANGLO-SAXON AGE John Blair
THE ANIMAL KINGDOM
 Peter Holland
ANIMAL RIGHTS David DeGrazia
THE ANTARCTIC Klaus Dodds
ANTISEMITISM Steven Beller
ANXIETY Daniel Freeman and
 Jason Freeman
THE APOCRYPHAL GOSPELS
 Paul Foster
ARCHAEOLOGY Paul Bahn
ARCHITECTURE Andrew Ballantyne
ARISTOCRACY William Doyle
ARISTOTLE Jonathan Barnes
ART HISTORY Dana Arnold
ART THEORY Cynthia Freeland
ASTROBIOLOGY David C. Catling
ATHEISM Julian Baggini
AUGUSTINE Henry Chadwick
AUSTRALIA Kenneth Morgan
AUTISM Uta Frith
THE AVANT GARDE David Cottington
THE AZTECS Davíd Carrasco
BACTERIA Sebastian G. B. Amyes
BARTHES Jonathan Culler
THE BEATS David Sterritt
BEAUTY Roger Scruton
BESTSELLERS John Sutherland

Available soon:

For more information visit our website

www.oup.com/vsi/

William H. Brock

THE HISTORY OF CHEMISTRY

A Very Short Introduction

OXFORD
UNIVERSITY PRESS

OXFORD

UNIVERSITY PRESS

Great Clarendon Street, Oxford, OX2 6DP,
United Kingdom

Oxford University Press is a department of the University of Oxford.
It furthers the University's objective of excellence in research, scholarship,
and education by publishing worldwide. Oxford is a registered trade mark of
Oxford University Press in the UK and in certain other countries

First edition published in 2016

Published in the United States of America by Oxford University Press
198 Madison Avenue, New York, NY 10016, United States of America

British Library Cataloguing in Publication Data
Data available

Library of Congress Control Number: 2015949588

ISBN 978-0-19-871648-8

Printed and bound by CPI Group (UK) Ltd, Croydon, CR0 4YY

Contents

Acknowledgements

Most of my lengthy *History of Chemistry*, published over two decades ago, was researched and written at the Chemical Heritage Foundation (CHF) in Philadelphia where I spent 1999–2000 as an Edelstein International Fellow. Considerable advances in the interpretation of the history of chemistry have occurred since then, and the present book has provided me with an opportunity to incorporate some of them.

In writing this very short introduction to the subject I am once again grateful to CHF for awarding me a Doan Fellowship in September 2014 which allowed me to use its excellent study facilities and to wallow among the books in its wonderful Othmer Library. This American sojourn gave me time to contemplate the problem Hugh Lofting's Dr Dolittle faced when filling his children's zoo: 'What to leave out and what to put in?' (*Dr Dolittle's Zoo*, 1925). I acknowledge the interest and help given me by the CHF library staff and fellow scholarship holders, especially Stefano Gattei (Lucca). I am grateful to Professor Hasok Chang (Cambridge) for suggesting that I should write this monograph. As always, my wife Elvina has been a patient and interested support to me.

List of illustrations

Introduction

Wer sie nicht kennte
Die Elemente,
Ihre Kraft und Eigenschaft,
Wäre kein Meister
Über die Geister
[Anyone who does not know about the elements, their
properties, their powers, will never be master of the spirits.]
(Goethe, *Faust*, Part 1, Scene III)

Chemistry has always been the science of matter and a technology
for creating new things through metamorphosis and
transmutation. According to recent statistics there are now more
than three million people in the world who call themselves
chemists and they publish over three quarters of a million
research papers annually. Most of these report the synthesis and
properties of new materials. In 1800 the elements and compounds
known to chemistry numbered only a few hundred; today, they
number more than seventy-one million, or roughly one substance
for every one hundred of the seven billion people now on earth.

Few of these substances, which include some hundred elements,
actually exist in nature; rather, they have been isolated, prepared,
and studied by chemists in particular times and places by an
evolving repertoire of laboratory practices and theoretical insights,

and recorded in publications in various languages. Only those chemicals that have been found useful have been given some kind of permanent existence. The rest remain as historical virtual curiosities, though they can be conjured into existence when required from the deep technical knowledge chemists possess of the constitutions, structures, and interrelationships between molecules.

Until William Whewell coined the word 'scientist' in 1834, those who devoted all, or part, of their lives to the study of the natural world were referred to as 'natural philosophers'. By the 17th century, however, specialization had begun and natural philosophy tended to refer to the more mathematical and quantitative interpretations of nature. Those involved in the study of plants and animals were said to practice natural history, and those studying the properties and reactions of different kinds of matter, and their exploitation to improve the human condition, were referred to as chemists.

As bodies of knowledge and practice the individual sciences have always enjoyed shifting boundaries. This is particularly true of chemistry which, until the 1840s, included most of what was later known as experimental physics. Although the scientific disciplines came to have fairly distinctive boundaries in the 19th century, as university departments, academic journals, and professional societies forged collective professional identities, chemistry never became an isolated discipline. Indeed, we have now reached the stage in the 21st century when cultural historians are asking whether it any longer makes sense to speak of chemistry as a separate discipline, because the study of stuff has become so interdisciplinary—with collaborative research and teaching involving mathematicians, physicists, biologists, and engineers.

For the purposes of this short introduction to the history of chemistry, however, we shall use the definition of chemistry proposed by the German organic chemist Friedrich Kekulé in 1861: 'Chemistry is the study of the material metamorphosis of

2

materials'. When Christian missionaries first began to translate western textbooks into Chinese in the 1870s, they needed a term to stand for *chemistry*. They coined the phrase *hua-hsüeh*, literally meaning 'the study of change'. This was an eminently sensible neologism, for chemistry is about change. As the American historian of chemistry Robert Siegfried noted, there is a magic about chemical change.

> Bodies disappear and new bodies with different qualities appear in their stead. A piece of metal can be added to a clear, colourless liquid, the metal disappears, and a blue colour appears. Based on direct experience alone there is no explanation available.

This book describes how explanations were found and new transmutations were discovered, studied, and exploited.

Chapter 1
On the nature of stuff

Early in our education we are introduced to the history of man in terms of stone, bronze, and iron ages—terms first introduced by the Danish archaeologist Christian Thomsen in 1836. However, the association of these terms with the development of chemistry is hardly ever made explicit.

At first sight, the term 'stone age' does not seem to imply anything chemical; but deeper reflection shows that it implies not merely the altering of stones to fashion tools and buildings some two million years ago, but recognition of different kinds of stuff (minerals) and their differing properties. Chemistry helped to define culture. The Palaeolithic (Old Stone Age) paintings of Lascaux and elsewhere show that stone-age peoples were able to prepare pigments to colour their representations of animals (12,000–8,000 BCE); and it would have been only a small step from this to the use of minerals and plants to extract natural dyestuffs with which to colour cloth and clothing once society had reached an agrarian stage of development.

By the time the bronze or copper age was reached, the working of minerals to produce copper and zinc had been developed. The archaeological evidence for mining and cupellation (the removal of lead in an ore by oxidation) is extensive, while the analysis of food remnants at metallurgical sites implies that the early metallurgists enjoyed a rich diet of meat and fish—suggesting that they were admired and respected for their ability to transform useless hunks of rock into valuable metal. These metallurgical practices continued with the discovery of iron ore, which gave iron-age man the ability to produce weapons that made animal hunting and butchering easier.

Archaeological chemistry combines analytical chemistry with the techniques of industrial archaeology to investigate sites of metallurgical activity. Chemists in the 18th century pioneered the investigation of ancient glassware and other objects, but since large samples were required for analysis, archaeologists were wary of deploying analysis on a large scale for fear of destroying evidence. This reluctance to compromise specimens was dramatically altered from the 1960s onwards, when new analytical instrumental techniques permitted the use of exceedingly small samples of a material object.

Artisanal science

Excavations in Egypt and the former lands of Persia and Mesopotamia have revealed the chemicals identified and exploited by these advanced civilizations from the fifth millennium onwards. For example, *natron*, an impure form of common salt extracted from dry lakes or from the evaporation of Nile water, was used in embalming and food preservation. Perfume recipes have scarcely changed for thousands of years. A basic ingredient has always been glycerol, a viscous sugar alcohol. We know from Pliny's *Natural History* (1st century CE) that there was an extensive Roman technology that involved the exploitation of the algae (*Dunaliella salina*) that is able to grow in the hostile environment of salt evaporation pans. Pliny and others referred to it as *flos salis* or 'flower of salt'. The cells of this brick-red halophile contain glycerol and beta-carotene (the precursor of vitamin A). Both these valuable commodities are still extracted from the algae today as the basis of coloured perfumes just as they were in Roman times.

We cannot be certain when mankind first tamed fire from lightning strikes and learned to kindle wood to make its reproduction readily available for the purposes of cooking and heating materials; but this was certainly a precondition for the production of pottery and glass (used in jewellery), and the extraction of metals. Trees and bushes burn to leave a charred mass (charcoal) which was found to release molten metals from rocks, stones, and minerals when they were burned together. Archaeologists have found hearths in sites dated at 230,000 years old. The findings imply that various hominid species had learned how to create, control, and propagate fire. Its control provided humans with warmth, expanded their daylight hours, made migration and a peripatetic lifestyle possible, and improved mankind's diet, health, and longevity through the processing of meat and vegetables. There is abundant archaeological evidence that by 8,000 BCE mankind was using biochemical processes (fermentation) to exploit grains of various kinds to bake bread and to create beer and wine.

The ability to control fire and temperature led to the first chemical technologies—the production of pottery from fired clays and tempers (such as sand and limestone), metals, glass, and bitumen products. The existence of exposed bitumen in Mesopotamia led to the creation of crude asphalt obtained by mixing the naturally found tar with chalk, sand, and gravels. Asphalt proved an efficient sealant and mortar for housing and boat building. Another early chemical technology was the production of plaster from gypsum or limestone that was then used to make large blocks for building purposes.

Glass and dyestuffs

Glass appears to have been developed from the initial glazing of stones of quartz using soda water. Such glass stones ('Egyptian faience') are found in sites dating from 4000 BCE. From these objects there developed a genuine glass made from sands (which contained iron salts as impurities), lime, soda, and heat dating from about 2500 BCE. By the 6th century BCE, antimony ores were being added to the glass fluxes in order to remove the green tinge of glass caused by its iron content. Clearly this represents a programme of empirical development of technology by trial and error of the kind familiar to the school children—let's see what happens when we heat *this* stuff with *that*. Applied chemistry (today's chemical engineering) is as old as mankind and was a sophisticated art long before a science of chemistry emerged.

Glass must have puzzled ancient technologists: was it a metal, since it was malleable when heated? Herodotus calls it 'molten stone' or water solidified by heat. Its astounding manufacture from such disparate stuffs as sand, alkali, and various oxides, and its astonishing ability to take up colour, may well have first suggested the concept of transmutation. If glass could metamorphose into valuable stones, might not metals generally be transmuted?

Among the ancient Egyptians, glassmaking was under the direct control of the Pharaohs and considered a holy profession. Indeed, glass, not gold, was revered as the noblest material. However, although glass could be moulded and coloured to make artificial jewellery, glass blowing, using hollow reeds tipped with a clay mouthpiece, appears not to have been introduced until the 1st century BCE, and only then was glass moulded into beakers and dishes. Glass working in the various areas of what were later described as the Holy Land had a continuous history that culminated with the emergence of Venetian glass workers in the 15th century. Glazed pottery followed from the discovery of glass and the further ability of glazes to take decorative pigments to enhance a pot's charm and value.

Such pigments were not merely obtained from crushed coloured rocks but also from heated mixtures that warrant the term 'synthetic' materials. For example, a blue pigment used extensively in Egyptian tomb painting was a complex of calcium, copper, and silicon oxides. Evidence collected from Persia and Babylonia has revealed even more sophisticated dyes and pigments.

Because of its stability, as much as for its striking colour, the most valued dye in antiquity was royal purple (dibromoindigotin) which was extracted from molluscs by boiling them in tin vessels. This produced a 'white' version of the dyestuff that was readily absorbed by cloth. On reoxidation in the air the full blue colour emerged—a striking transmutation. A virtually identical process seems to have been used by the woad dyers of ancient Britain. Woad leaves were crushed and rolled into balls and left to ferment. The balls were traded in this form. A second fermentation in warm water produced a soluble form of indigo. A cloth immersed in this liquid oxidized in air to the insoluble form of indigo which was fixed into the textile fibres. The technology of dyeing cloth was to remain little changed until the 18th century.

Metallurgy

While such usages of 'manufactured' chemicals signalled the ability of natural stuffs to change their identities, appearances, and properties, it was the extraction of metals from rocks and quarries that advanced these early examples of chemical technology the most—as the terms bronze and iron ages imply. In this respect, it is significant that the ores from which the common metals copper, iron, tin, silver, and lead were first obtained by oxidation and reduction in a smelting process were all (with the exception of tin) used as pigments before they became the subject of metallurgical exploration. The oxides and sulphides of arsenic and antimony (realgar, orpiment and stibnite) were used as pigments, though the metals themselves were not separated until the 13th and 16th centuries BCE. The red oxide of iron was used in the decoration of walls, while copper minerals were used as cosmetics long before iron and copper became objects of chemical manipulation. Of course, many metals such as gold, electrum (an alloy of silver and gold), and copper existed naturally, but scarcely in such abundance as to account for their widespread use.

It would appear that the earliest mines and metallurgical operations in the West took place in Mesopotamia, and that metals were traded for use in Egypt. Archaeologists are not certain how the necessary conditions for efficient smelting of the common metals was achieved, though it seems probable that it was a result of techniques that had first been applied in pottery making—that is, using a kiln. Indeed, the common feature of ancient technology is the kiln, which could serve both as an oven for cooking meat and baking bread, but was also adaptable to hardening clay that had been shaped into pottery, and further adapted by controlling the flow of heat to make glass, or to extract metals from their ores. By yet further modification the kiln was adapted into a subliming and distillation vessel.

As we shall see, from time to time, apparatus, instruments, and experimental techniques have played a crucial role in accelerating the progress of chemical knowledge or in pushing it in new directions. The kiln can be regarded as the oldest piece of chemical apparatus and instrumentation in that it (and its adaptations) propelled the advancement of metallurgy and the discovery of alcohol and mineral acids.

Recipe literature

In a sermon by one of the Early Church Fathers (Theodoret of Cyrus), chemistry was invoked to justify belief that the Divine Word had become flesh: sand became glass, grapes produced wine, and wine became vinegar. In each transformation a new stuff appeared and received a different name, and in each case something superior and useful was created.

Ancient civilizations had knowledge of seven metals (gold, silver, copper, lead, tin, iron, and mercury) and a wide variety of 'chemicals' that they exploited in their pottery, jewellery, cosmetics, cooking, and weaponry, and as drugs. Our knowledge comes from literary evidence in the form of Babylonian cuneiform tablets and Egyptian papyri, and the many thousands of pre-1500 CE manuscripts that record recipes for making pigments, paints, and inks.

As the Smithsonian Institution museum curator Robert Multhauf concluded, 'practical chemistry appears to have changed very little during the twelve centuries between the papyrus Ebers (c.1550 BCE) and the oldest extant Greek treatise' on the subject. He noted, however, that this did not mean that development was non-existent. Progress was simply difficult and made up by the accruement of small steps. Using later chemical knowledge, it is clear that these ancient chemists had to master a large number of practical steps, most notably the thermal control of chemical processes, to produce metals and minerals in a state of comparative purity, or at least of more or less constant composition and hence reliability.

The recipes in the so-called Stockholm papyrus, which dates from around 300 CE, tend to be either concerned with debasing gold to increase its bulk by adding copper, or colouring the surface of other metals to make them look like gold. How did the ancients judge whether such gold was genuine or spurious, or impure? At least three tests were known to the Egyptians and were employed well into the later Middle Ages in the Latin West. In the simplest test, a touchstone, a black abrasive stone, was scored against a sample. The 'artificial' gold streak obtained on the touchstone was then compared with the colour of a score obtained from a pure sample. A more sophisticated and non-damaging test was to take comparative specific gravities of the test and pure samples. This is the test that supposedly led to Archimedes' 'eureka' moment in the bathtub, but literary and pictorial evidence shows that it was known long before the time of Archimedes. A third, truly chemical, test was by the use of fire. If the gold sample was impure, when heated sufficiently strongly any impurities were oxidized and discoloured the crucible in which the test or assay was made. Clay crucibles may have been the first pieces of chemical apparatus invented, and date from the 6th millennium.

Egyptian and Greek dyers, glassmakers, metallurgists, jewellers, and pharmacists are known to have engaged in tincturing and colouring metallic surfaces from at least the 4th century BCE onwards. Such procedures, some of whose recipes have come down to us, were not necessarily fraudulent in practice or intent, but were probably equivalent to the modern production of cheaper synthetic materials. Joseph Needham, the Cambridge biochemist and historian of Chinese alchemy, described this technology as *aurifiction*. Fraud, fakery, and deliberate deceit were only too possible, and much of the rich and entertaining literature on alchemy, such as Ben Jonson's play *The Alchemist* (1610), hinges on such trickery, human frailty, and greed.

Why do stuffs change appearance?

The 19th-century physicist John Tyndall noted in his journal, with some degree of puzzlement, that his laboratory assistant, while dextrous in designing and handling equipment, had no interest in why experiments worked or what explained physical phenomena. It appears that no fundamental science was involved in the artisanal practices of the ancients; indeed, almost all 'industrial' or 'applied' chemistry right up until the 18th century was purely 'trial and error' and empirical in nature. Today we would call this an engineering approach—seeing what an object does rather than explaining why it does it.

This apparent indifference to what we regard as necessary scientific explanation does not mean that artisans had no cultural notions about what they were doing and witnessing. An Assyrian cuneiform text dating from the 8th century BCE, for example, implies that rituals accompanied metallurgy, and this is confirmed by the long historical association between the common metals and the seven heavenly bodies known to ancient astronomers (e.g. the Sun with gold, the Moon with silver, Mars with iron). That in turn shows that early mankind associated heavenly phenomena with earthly events. Such 'astrological' links, and more generally beliefs in the correspondence between the macrocosm and microcosm, remained a fundamental part of chemistry until the 17th century. However, by the 3rd century BCE a few lone thinkers in the Mediterranean, made rich by their trading in the region, had begun to proffer explanations for transmutations and changes applicable to both the natural world of the microcosm and to those conducted by artisans in their workshops.

Speculations about matter

No theory was required by artisans in the empirical skills of perfume making, glassmaking, cosmetics making, pottery making, bronze making, or the creation of gold or gold-like objects. But the Greeks demanded to know *why* and *how* such transmutations of form (what we call chemical change) were possible. Why did metals or glass change appearance when melted with coloured minerals? Why did beads of silver appear when galena (impure lead oxide) was heated?

Faced by such apparently realistic and successful examples of metamorphosis by artisans, Greek philosophers saw these as real transmutations similar to the change of water into air by evaporation, or the growth of an acorn into an oak tree: that is, changes of essence and not of mere appearances. The Greeks were thinkers, and asked themselves what matter was. How and why does it change in appearance and properties?

The earliest thinkers, known as the Presocratics, suggested that change was an illusion and that matter was ultimately just one material. Thales (*c.*625–*c.*547 BCE) famously suggested that the

Urstoff, or primary matter, was water; others suggested air or fire. Empedocles (*c.*494 BCE) proposed that there were four basic materials: earth, air, fire, and water. Other suggestions were that matter was infinitely divisible; or that division was limited until one came eventually to indivisible atoms that differed only in shape and size and their rapid movements in a void.

Aristotle (384–322 BCE), the most creative of these early philosophers, laid stress upon the sensual qualities of matter, its hotness or coldness, its wetness or dryness, and attached these qualities to the four elements of Empedocles (see Figure 1). None of the elements was unchangeable. The element water was obviously cold and moist, but when heated it transformed into air whose qualities were hot and dry. These elements and their qualities made up the composition of the metals, minerals, and plants of the observable microcosm. For centuries the four elements were believed to be the roots of all matter, and even to the present day they remain embedded in chemical terms such as alkaline earths, rare earths, fresh (pure) air, inflammable air, *aqua fortis*, and *aqua regis*.

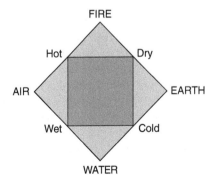

1. **The four elements and their qualities.**

Aristotelian chemistry

In the hands of Aristotelian philosophers, the possible transmutation of one metal into another was theoretically justified as the transfer of forms and qualities that differentiated an otherwise basic, formless, underlying matter. Later philosophers called this *hylomorphism*.

Aristotle's categories and matter theory more or less obliterated those of his teacher Plato (427–348 BCE), whose *Timaeus* had postulated a structural and mathematical model of matter based on triangular and polyhedral forms. This theory, too, had sensual and possibly technological roots, based upon the shapes and textures of observable materials. However, such structural, corpuscular, or atomic speculations were overwhelmed by the more powerful Aristotelian model and only reappeared in the Latin West in the late Middle Ages.

Equally influential was Aristotle's analysis of causation: what is a particular stuff's purpose (a final cause)? What makes a stuff have its particular shape, appearance, and properties (formal cause)? What is it made from (material cause)? And what causes it to express the properties that it does (efficient cause)? In the case of a metal, for example, its purpose can be for construction, its form or properties are given by it being a metal derived from its particular qualities, its material nature is given by its composition from earth and water, and its efficient cause is to be found in its long growth inside the earth from a trapped dry smoky exhalation that coupled with another moist vaporous exhalation. Other non-metallic minerals were formed in the same way depending on the relative amounts of the two exhalations. In the hands of later Arab chemists this explanation was developed into the sulphur-mercury model of composition.

Alchemy

The earliest surviving Greek text concerning alchemy is 'Of natural and hidden things' (*Physika kai Mystika*). In it we find for the first time the division of the great work of making gold (*chrysopoeia*) into the four experimental stages of blackening (*nigredo*), whitening (*albedo*), yellowing (*citrinatus*), and reddening (*rubedo*). The book is often attributed to Democritus, possibly the atomist. Transmutations are mainly done the dry way, that is by fire.

The development of the water bath, or *bain marie* as it is still known in Europe, is attributed by Zosimus of Panopolis (fl. 300 CE), the greatest authority among Hellenistic alchemists, to a female alchemist named Mary the Jewess. We know nothing about where she lived, or when she was active, though from her references to God as having directly revealed the secrets of nature to her and that they were not to be passed on to those who were not of 'the race of Abraham', her Jewishness is confirmed.

Zosimus quoted extensively from one of her manuscripts entitled *On Furnaces and Apparatus*, and it is from this that we know she constructed and used some eighty devices made from glass, clay, and metals to sublimate and distil various materials like mercury, which she described as a 'deadly poison, since it dissolves gold, and the most injurious of metals'. Among her appliances, no doubt derived from cookery, was a double vessel, the outer one of which was filled with water and heated while the inner vessel contained the materials to be treated by the gentle warmth.

It seems quite probable that, like Liebig's 19th-century condenser, the water bath had been in use for centuries before and that it was the fame and prestige of Mary's writings that led to it being forever associated with her name. Whatever the truth of the association, her book on apparatus, which also included accounts of various forms of still, influenced later Arabic alchemists and through them, practical alchemy in Europe.

Zosimus and other Greek alchemists collected many alchemical recipes from Mary including ones for the preparation of the philosophers' stone, a mysterious seed that supposedly precipitated an instantaneous transmutation. These instructions, while clearly based upon real chemical manipulations by heat or distillation, were already written in a figurative language that relied upon the appreciation of analogies between metals and man, and matter and spirit.

Chemical language

The ancient (al)chemists obviously derived much of their apparatus and manipulative techniques from the equipment used and developed by artisans, technologists, metallurgists, and pharmacists, whose heating, cooking, subliming, and distillation techniques were grist to the mill of later chemists.

Alchemy also provided later chemistry with the idea of a symbolic language for practitioners of the art. Here, however, the multiplicity of synonyms for the same thing served not merely to obscure matters for the uninitiated but to increase greatly the degree of symbolic allusion—to the utter mystification of later readers and interpreters. For example, according to one Greek alchemical lexicon, mercury was 'the seed of the dragon', or 'dew', or 'milk of the black cow', or 'Scythian water', or 'water of silver', 'water of the moon', 'river water', and 'divine water'. A Persian source lists over fifty synonyms for the philosophers' stone.

Given that the same synonyms were also used for different substances, the resulting poetical muddle must have been as confusing for alchemical practitioners as it is for later historians.

Isaac Newton (1642–1727), for example, spent years compiling an *index chemicus* in an attempt to make sense of alchemical language and allegory. Why the secrecy and obfuscation of language? We have to recognize that chemistry as late as the 17th century was still not a public science and that in the absence of a patent system, methods for producing medicines, or carrying out chemical procedures that had potential cash value in a world of saleable commodities, were best kept secret or only shared among cognoscenti. Moreover, gold making, if ever perfected, threatened state economies, or it might be used to usher in the millennium as some religious fanatics believed.

Historians have become adept at understanding obscure alchemical texts with their extraordinary illustrations and multiple hidden names (*Decknamen*, as scholars call them) for common substances. It turns out that these can be read like cryptic crosswords as accurate accounts of physical and chemical changes by reproducing alchemical recipes in the modern laboratory. Chemistry was to be a rich source of metaphors throughout its history, and it seems chemical metaphors were first adopted by bards who knew their audiences were familiar with the blacksmith's workshop.

Arabic alchemy

The development of the monotheistic religion of Islam in the 7th to 10th centuries led to a period of profound learning in the Muslim empire that embraced all the great centres of former Greek and Roman culture. The word 'alchemy', meaning 'the art of chemistry', is a reminder of the important contributions that the Arabs made to chemistry. Several other Arabic words such as alkali, alembic, elixir, and alcohol remain part of our English vocabulary.

Chemical and pharmaceutical recipes and apparatus were inherited by the Arabs and developed by, among others, Jābir ibn

Hayyān (*c.*721–815). We know little of his life, but it is certain that a Jābirian school of chemists emerged that knew and taught how to prepare mineral acids and alcohol through intricate distillations. The generation of metals in the ground was now attributed to interaction between highly purified and ethereal versions of mercury and sulphur. Even more significantly, Muslim alchemists developed the model that different metals contained internal and external qualities. Thus, if gold was hot and wet externally, it was cold and dry internally; whereas silver was the opposite. Hence to transform silver into gold, a reversal of qualities was needed.

Arabic and Syriac theoretical and practical knowledge passed to the Latin West via a variety of geographically different translation routes. By the 13th century we find many of the data accumulated by Arabic chemists summarized in the *Summa Perfectionis* by a Franciscan friar named Paul of Taranto, who used the pseudonym of 'Geber' (see Figure 2).

The relationship between the writings of Jābir and Geber caused much controversy between German and British scholars in the early 20th century. We now believe that the writings attributed to Geber, while based on Arabic originals, included many original additions. The most important of these, according to the detective work of the American scholar William Newman, was a form of corpuscularianism derived from Aristotle's concession that (despite his objections to atomism) there were *minima naturalis*, or 'molecules' as we would say, which limited the analytical division of matter.

In Geber's writings, Jābir's internal and external qualities were interpreted in corpuscular terms. This gave new power to the notion of transmutation whereby one searched for a tool that would lead to a structural rearrangement. One tool suggested was mercury. But the Geberian school of alchemy was also

GEBERI PHILOSOPHI AC ALCHIMISTAE

MAXIMI, DE ALCHIMIA
LIBRI TRES.

Eiusdem liber inuestigationis perfecti magisterij , artis Alchimicæ.

Iis additus liber trium verborum.

Epistola item Alexandri imperatoris , qui primus regnauit in Græcia,
Persarum quoq; extitit imperator : Super eadem re.

2. The title page of *Philosophi Ac Alchemistae* (1529) attributed to
Geber showing two chymists hard at work behind a distillation vessel.

thoroughly Christian, so transmutations were also seen as metaphors for spiritual transformation, improvements, and rebirth.

What's in a name?

The presumption of early 20th-century historians was always that there have been obvious demarcation lines between chemistry and alchemy, and that these were clear even before the 16th century. Our use of two different words implies historiographically that the two fields were distinguished in the past—that there were two distinct disciplines. Chemistry was mechanistic, alchemy was vitalistic and spiritualistic. This interpretation is not supported by recent studies, which show that alchemists often thought in terms of minimal particles or corpuscles, and that they were interested in the quantities and weights of the stuffs they reacted together just like real chemists.

The American chemist historian Lawrence M. Principe, working with William Newman, has emphasized that any attempt at differentiation involves using contemporary distinctions backwards. In the 17th century the terms alchemy and chemistry were used interchangeably. Alchemy restricted to gold making (chrysopoeia) is a later distinction. For example, we find books entitled *New Light on Chymistry* (1604) that consist entirely of transmutation recipes (i.e. it's alchemical), while an influential book by Andreus Libavius (1560–1616) is entitled *Alchemia* (1597) although it is not about gold at all, but concerned with the chemical operations of distillation and crystallization, and the preparation of pharmaceuticals (see Figure 3).

To avoid confusion, Principe and Newman have suggested that historians should use the archaism *chymistry* when referring to early-modern chemical *and* alchemical practices before the end of the 17th century. The current consensus is that chemistry grew directly from what was called alchemy or alchymia. If we

3. Title page of Libavius' *Alchymia* **(1606). This folio volume contains about 200 illustrations of chemical glassware, apparatus, and furnaces, as well as designs for an ideal chemical laboratory.**

consistently apply the term chymistry to all chemical practice before the 18th century we also underline the fact that although the pre-modern world view was quite different from those held by 21st-century chemists, its practical activities were those that are familiar today. But if so, *how* did the terms alchemy and chemistry become differentiated?

Chymistry becomes chemistry

In *De la Pirotechnica* (1540) the Italian technologist Vannoccio Biringuccio summarized existing knowledge on mining, metal, and glass working—knowledge that had long been protected by the secrecy of the artisans' guilds, though frequently 'leaked' by the many 'books of secrets' that had circulated since the appearance of *Secretum secretorum* in the 12th century.

Biringuccio was one of a rising band of 'artist engineers' who began to publish the knowledge they had gained from their practical artisan workmanship, and to compare this knowledge with the Greek classical heritage and the theories of matter of the chymists. The result was a critical appraisal of the theoretical with the observational. In particular, Biringuccio wanted to free alchemy from its overlay of Hermeticism and to develop its quantitative aspects in order to improve metallurgical techniques, especially for the production of artillery. Another important handbook of the same ilk was Geog Agricola's *De Re Metallica*, published a decade after *De la Pirotechnia*.

It was only in the 18th century that the term 'alchemy' became restricted to the futile task of the transmutation of base metals and 'chemistry' to the analysis and synthesis of materials. Even then we find that 'chemistry' was largely restricted to a medical context inherited from chemiatria or iatrochemistry (Chapter 2). It is in this century that we see alchemists dismissed as frauds and alchemy as a hopeless cause. By the time we reach the great French *Encylopaedia* of 1753, chemistry is defined as the science 'which concerns separations and unifications of the principles making up bodies', and alchemy as 'the art of transmuting metals'. In other words, the synonyms alchemy and chemistry have become non-synonymous.

The disappearance of alchemy

The fundamental reason for the disappearance of alchemy (in the sense of chrysopoeia) was that it did not work; what did work, however, was that chymists could earn real money by being useful in helping a nation's economy and by preparing for sale various chymical nostrums.

The German chymist Johann Glauber, for example, made his fortune not from making gold but by preparing and selling 'sal mirabelle', a wonderful salt he had accidentally prepared when forming hydrochloric acid by the action of sulphuric acid on table salt. Glauber's salt, as it became known, and we know as sodium sulphate, proved a highly efficient laxative. Another interesting alchemical figure, Leonhard Thurneysser, became a rich man when, under the patronage of the Elector of Brandenberg, he established a factory employing some 300 people for the production of saltpetre, acids, alums used as mordants, glass, and pharmaceuticals.

In the last decade, historians have traced the downgrading of alchemy into a pseudo-science. This boundary work was largely done by French chemists working in the Académie des Sciences from the 1740s onwards. It was achieved by delimiting alchemy to gold making (whereas, as we have seen, it was rather more than that) and dismissing alchemists as greed-driven frauds.

Alchemical facts were also taken out of their alchemical contexts, reinterpreted, and transformed into reproducible chemical facts. For example, consider this impressive and wonderful alchemical experiment. If a mixture of nitric acid, mercury, and silver are allowed to stand for a couple of months, the mix 'vegetates' into a tree-like form of silver called Diana's Tree. But whereas alchemists had seen this tree as evidence of a vital spirit and a key step en route to the philosophers' stone, French chemists discussed it under the properties of silver and as an example of mechanical crystal formation. They reinforced this interpretation by showing how similar trees could be grown using lead or iron instead of mercury.

We should note an important historiographical point here. The alchemists had not lied about such trees: although the transmutation of base metals into gold was a chimera, alchemists' accounts of the chemical changes they witnessed were not imaginary.

French chemists still had to concede alchemy's theoretical possibility because of their own growing commitment to corpuscular models of chemical change, but they diminished its significance by only mentioning it under the rubric of the chemistry of gold and not as a subject in its own right. French academicians had the power to enforce the separation and distinction because its members, under Royal prerogative, acted as reviewers for publications. So any gold-making claims could be traduced and explained away by ordinary acid-alkali chemistry. Moreover, the academicians extended this critical reviewing and boundary work by reinterpreting older reports of transmutations as chemical facts.

Alchemy's heritage

Alchemy, and chymistry more generally, bequeathed to modern chemistry a rich variety of chemical operations, manipulations, techniques, and apparatus, but not the conceptual frame of modern chemistry. Historians can explain alchemy's demise by its flaws—mysticism, metaphoric language, and its failure to work. What did work was its application to useful arts and, above all, its service to pharmaceutical practice. In such areas commercial gains were to be found that were noticeably absent in the attempt to transform base metals into gold. The alchemists' primitive ideas of corpuscular composition also proved a useful theoretical thread that led towards modern chemistry.

From the 1960s onwards, historians have increasingly viewed alchemy in its cultural and historical context, and it is here we find that its practitioners made no distinction between gold making and practical chymistry. Rulers in the European courts supported both practices while scientific icons like Boyle and Newton practised both.

Chapter 2
The analysis of stuff

The fundamental problem in chemistry is transmutation. How can two homogeneous stuffs with very different properties merge to form another homogeneous material whose properties are different from the reactants? Since the reaction can be reversed and the original reactants regenerated, it would seem that they are in some weird way still present in the product. But if so, why are the properties completely different and not merely an averaging of the reactants' properties?

Aristotle had met the issue of homogeneity by teaching that the reactants' forms were present in the product 'potentially'. This was not just a chemical problem; it was one that medieval theologians wrestled with because of the issue of the Catholic teaching of transubstantiation of bread and water into the body of Christ. Consequently, any rejection of Aristotelian philosophy could be regarded as a rejection of a fundamental Christian doctrine.

The rediscovery and publication of Lucretius' didactic poem, *De rerum natura* (*On the Nature of Things*), in 1473 brought Greek atomism back into circulation. In principle, atomism gave a more satisfactory explanation of the homogeneity problem. Atoms came in various shapes and sizes and different substances arose from the various structures the atoms assumed.

The theory's fatal weakness (apart from its atheistic implications) was that there were no rules that explained why atoms adhered to each other in particular ways other than unsatisfactory appeals to attraction and repulsion. Even so, we find a general revival of corpuscular explanations among chymists during the 16th and 17th centuries; but their corpuscles were not featureless atoms, but property-bearing particles that showed a compromise between original Greek atomism and Aristotelian forms. Even in the 21st century, the problem of homogeneity has to be explained by additional concepts that go beyond bare atoms: ideas of valence, bonding, electronic shells, and orbitals.

We have seen how the origins of chymistry lay in the metallurgical, distillation, and glassmaking activities of Graeco-Egyptian artisans. Knowledge of these practices was passed on to the Arabic conquerors of Europe from the 9th century onwards, and Arab chymists were able to isolate mineral acids such as nitric and hydrochloric acids, as well as alcohol. When knowledge of Greek and Arabic chymistry passed to the Latin West through translation and trade, these new distillates were gradually put to use in the preparation of medicinal materials by, among others, the Franciscan John of Rupescissa (fl. 1340), whose preparation of 'quintessences' proved useful to Paracelsus (1493–1541) and his followers (see Figure 4).

Paracelsus not only expanded the Arabic doctrine that two principles, sulphur and mercury, were the roots of all things by adding a third principle, salt, but he taught that the universe itself functioned like a chemical laboratory. God the Creator, he believed, was a divine alchemist whose macrocosmic drama was mirrored in the microcosmic world of man and earthly creatures. It followed that physiological and pathological processes were chemical in nature, and that disease was best treated by chemical medicines rather than by the herbal ones of the ancients.

4. Furnace with hooded distillation apparatus. Hieronymus Brunschwig, *Liber de arte distillandi de compositis* (Strassburg, 1512). This was the first printed book to describe the preparation of distilled medicines (quintessences) from plants and other materials.

Paracelsus and iatrochemistry

One of the most curious personalities of the 16th century was
Theophrastus von Hohenheim who was born in Switzerland in
about 1493, the son of the town doctor of Einsiedeln in the
German-speaking Swabian canton. His childhood was disturbed
by the Swabian war in which Swiss cantons struggled for
independence against the authority of the Holy Roman Empire,
and the family sought refuge in Austria in 1502. As a child he
suffered from rickets, but his father instructed him in medicine
and as a young man he also learned a good deal by talking to
miners and metallurgists.

Paracelsus' life was that of a wanderer. The only settled period in
his life came 1527 when, as a result of impressing some influential
humanists with his cures, he was appointed municipal doctor in
Basel. The job also included the inspection of the apothecaries'
shops, and he was soon accusing apothecaries of cheating their
customers. He also offended local physicians by attacking what he
saw as outmoded and incorrect Galenic medicine, lecturing in
German instead of Latin, and dressing in workmen's clothing
rather than those of a gentleman physician. A year after his
appointment he was forced to leave Basel and resume his life as a
wanderer, preaching and practising medicine, and printing his
tracts and homilies as the opportunity arose. He died in Salzburg
in 1541 and was buried in an almshouse (see Figure 5).

For publication purposes, Theophrastus took the Latinized name
of Paracelsus, which has often been taken to mean 'greater than
Celsus', the Roman medical authority of the 1st century CE; but it
was more likely simply an invented Latinization of his family
name, Hohenheim. His writings were filled with angry diatribes
against the Roman Catholic Church and at orthodox medicine. The
original heterodox 'angry young man', he welcomed confrontation
and avoided compromise. To come to terms with him in the

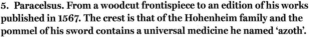

5. Paracelsus. From a woodcut frontispiece to an edition of his works published in 1567. The crest is that of the Hohenheim family and the pommel of his sword contains a universal medicine he named 'azoth'.

21st century is difficult, until one realizes that in the 16th century religion and science were inseparably two sides of the same coin. His medical and chemical work were but minor aspects of his religious beliefs and goals for the complete reformation of society.

In essence, Paracelsus was a 'spiritualist' who saw spirits from the stars and heavens as the essences of chymical bodies. He saw his mission in life to reform both natural philosophy and the theology of which it was part and parcel. Living as he did in the turbulence of Protestant attacks on the Pope's authority, and amid the disturbances of the Peasants' War against landlords and authority, and everyone's fears of a Moslem invasion of Europe from Turkey, Paracelsus agitated for wholesale revolution in religious practice and natural philosophy, believing that the millennium was at hand and that God had ordained him to be an itinerant prophet to prepare for the second coming of Christ.

It was in this deeply religious context that Paracelsus made his medical and chemical innovations. His reputation as a doctor probably lay in his conservatism and non-interventionalism on the grounds that nature can perform its own cures; to this he added a repertoire of medications—small doses of chemical remedies. He saw alchemy not as gold making, but as a way of separating useful objects from the useless through the extraction of their essences. He derided his contemporaries' adherence to Galen's humoral theory that assumed that disease was due to the upset of the normal balance of the four humours—blood, phlegm, and yellow and black bile—that made up a patient's normal constitution, and he argued instead that diseases were specific and had specific locations. He sought the causes of disease in inappropriate diet and from 'poisons' that were ultimately drawn from the stars.

Remedies could be identified from their 'signatures', using signs drawn from shape and colour in plants that resembled or suggested analogies with specific anatomical organs. For example, a yellow flower might indicate that it was suited to the treatment of a diseased liver. Existing herbal medicine often used similar arguments, but Paracelsus differed from herbal practitioners in not using the flower or herb itself as a medication, but in using an extract produced from the flower by chemical means such as distillation in alcohol. In addition, using his chemical and metallurgical knowledge, he prepared and used small quantities of metallic salts and distillates.

He thereby set in motion what became known as iatrochemistry or medical chemistry. This was to have a revolutionary effect in that once chemical therapies were accepted, chemistry had to become part of a physician's training at university. Alchemy or, more generally, chymistry, had never been accepted as part of the university curriculum; academic chemistry was now to develop in a medical context for the next two centuries.

Paracelsus' own religious framework was a form of devout Trinitarianism, and this evidently implied to him that matter and disease had a tripartite form and occurrence. Consequently, he held that three principles constituted the stuffs of the microcosm. Drawing on the Arabic and medieval material theory that metals were made from two principles, sulphur and mercury, he added a third material body, salt. He cited the way metallic lead took on very different tripartite physical and chemical forms according to whether it was minium (red lead oxide), ceruse (white lead carbonate), or a lead glass.

These transmutations could not be explained by a four-element theory, he claimed, but were easily explained in terms of matter's trinity of sulphur, mercury, and salt. The burning of a stick illustrated this model. The combustion was due to sulphur, the smoke was caused by mercury, and the remaining ash was salt. In general, salt conferred solidity and taste; sulphur gave colour, smell, and flammability; and mercury conferred volatility and metallicity. This was the doctrine of the *tria prima*, the three principles.

It may well be asked why such an extraordinary man always figures in the history of chemistry? The answer, besides the fact that his life vividly illustrates the intimate connection between science and religion in the early-modern period, is that his many followers promulgated a medical chemistry. Although there were bitter disputes between physicians who stuck to the administration of Galenic plant products as medicaments, and the Paracelsians who recommended the use of chemically prepared medicines and (bio)chemical interpretations of physiology, the Paracelsians won the war.

The ramifications of this iatrochemical movement were far-reaching. For the Paracelsians, chymistry was to be the handmaiden of medicine. Most notably, it meant that chemistry, or at least pharmaceutical chemistry, became part of the European medical curriculum and established in universities. This was something that alchemy (a technology) had never achieved in the Middle Ages. This was to prove of mutual benefit both for the advancement of medicine and for the emergence of academic chemistry.

The first professor of medical chemistry was appointed at the University of Marburg in 1609, and by the end of the 17th century chemistry was firmly established in medical faculties throughout Europe. To aid the instruction of medical students, more and more chemical textbooks were published, the most significant being by the Dutch polymath, Herman Boerhaave (1668–1738) at the University of Leiden. Paracelsus also influenced a range of significant figures in chemistry such as Helmont and Boyle, who used his doctrines as a foil in demonstrating their own beliefs; and, finally, the doctrine of the *tria prima* had a definite role to play in the development of ideas about composition and combustion. Paracelsus was the trigger for the start of the chemical revolution.

Van Helmont's water

Chymistry remained in a deeply religious context for the Flemish follower of Paracelsus named Joan Baptista van Helmont (1579–1644) who lived and worked in Brussels all of his life apart from travels in Europe as a young man. As a Flemish nationalist he had a deep resentment towards the Jesuits and other Schoolmen who occupied positions of educational and church authority in the Spanish-occupied Lowlands. Like Paracelsus he wanted to reform medicine, but abandoned practising it in 1609 to devote himself to empirical chymical research. His writings led to his investigation by the Spanish Inquisition and, like Galileo, he was forced to acknowledge his errors in 1630. He was condemned again three years later and spent some years under house arrest for promulgating 'the monstrous superstitions' of Paracelsus.

His major work, *Ortus Medicinae* (*Origins of Medicine*, but anglicized in the English translation of 1662 as *Physick Refined or Reformed*) was published posthumously by his son in 1648 and influenced the work of Robert Boyle in England and Boerhaave in the Netherlands. Helmont's cosmology was as grotesque as Paracelsus' and we are always in danger of picking and choosing the seemingly 'correct' bits.

For example, despite his having the idea of a primary matter that generates the *tria prima*, and although he sometimes wrote in terms of atoms and the void, all this was contextualized within an essentially Aristotelian framework that replaced forms with ferments, seeds, spirits, and strange emanations like Gas, Blas (a principle governing stellar movements and causing the tides on earth), and Magnall (a principle of light), which functioned essentially to convey or explain properties and qualities to individual substances and diseases. Such 'seeds' or 'ferments' were the necessary agents of transmutation of reactants into products.

Helmont was much more empirical and experimental in his investigations of nature than Paracelsus. He made extensive use of the balance, and careful observations during his experiments that were clearly based on a belief that matter was conserved during chemical changes. For example, he carefully analysed the different exhalations or smokes that evolved when different substances were burned (attributed to the element air by Aristotelians, and to the mercury principle by Paracelsians). He termed these smokes 'gas' because of their chaotic character in frequently exploding when collected in a closed vessel. He was even able to differentiate many of them.

But what was the significance of gas for Helmont? Because it was an exhalation he believed it was the pure essence of a substance, its spirit in fact, or what he termed its *archaeus*, the material manifestation of the substance. Matter and spirit were, for Helmont, two aspects of the same thing—unlike the dualistic philosophy of matter and spirit that was being popularized by René Descartes, a younger contemporary in the Netherlands about the time of Helmont's death. The term 'gas' soon disappeared since it was not taken up by other writers who preferred to use the terminology of 'airs'. We owe its revival to a French dictionary compiler in 1766.

Helmont denied that chemical reactions were real transmutations. When metals were dissolved in acids and apparently formed a watery solution, the metal could be recovered weight for weight. The original quantity of a metal could also be recovered when one metal precipitated another from a solution, as when iron precipitated copper from a vitriol solution of copper. His most famous quantitative experiment concerned the growth of a willow tree, from which he deduced that that all matter was ultimately water. He began with a sapling weighing 5 lbs which he planted in 200 lbs of soil. He watered the tree regularly for five years, following which the tree had increased in size and weight to 169 lbs, while the soil had neither lost nor gained any weight.

A modern chemist sees this as an undetermined experiment because Helmont had neglected the possibility that the tree absorbs nutrients from outside the soil in which it was embedded, namely from the atmosphere; otherwise his conclusion that the tree had grown solely from water seems impeccable. He was critical of both the four-element and three-principle theories, and argued instead for there being only one primeval element, water, though in practice he often combined this with air as if that too were elementary.

Daniel Sennert's corpuscles

The defenders of Paracelsian and Helmontian chymistry and medicine included many physicians who found patronage in the courts of central Europe, and who thereby acquired authority over and protection from the rival, critical Galenists outside such aristocratic protection. Others found jobs in the medical faculties of European universities.

An important iatrochemist was Daniel Sennert (1572–1637), a professor of medicine at the University of Wittenberg. He found a median way that reconciled the cosmologies of Aristotle, Galen, and Paracelsus with a corpuscular tradition that had long lain underground in the writings of various medieval chymists and which can ultimately be traced back to a theory of *minima naturalis*. Sennert taught that the different properties of substances were not due to the elements or principles of Aristotle or Paracelsus, but to a set of particulate seminal (formative) principles that produced new and different properties after the comingling and separation of reactants and combinations.

He backed up this primitive corpuscular theory with experimental observations. For example, since the vapour derived from distilling spirit of wine penetrated a wodge of writing paper, the vapour's particles must be very small; the phenomenon of sublimation could also be simply explained as due to the motion of particles. When gold and silver were mixed to form an alloy, the gold and silver had not disappeared but remained; the proof lay in the fact that they could be regenerated by adding *aqua fortis* (nitric acid) to the alloy. Such commingling and separability suggested particulate compositions.

We find the same arguments used by Boyle later, often copied word for word without acknowledgement to Sennert. Substances were composed from 'that into which they are decomposed' and vice versa. In such processes, matter was resolved into the *minima* (the smallest particles of the substance), which were then rearranged to form a new body whose properties resulted from a seminal principle (or form) particular to the new substance. Sennert called such reactions of separation and assembly *diacrisis* and *syncrisis*, and we can see in this the fundamental notion of analysis and synthesis. The *minima* were not necessarily indivisible atoms (a doctrine that could lead to accusations of atheism), but the smallest identifiable parts of a substance that resisted any further decomposition by any known chymical operations.

Sennert's corpuscularianism was inherited from an alchemical tradition that can be traced back to the 13th century in the work of Geber in the *Summa Perfectionibus*, a work that drew upon Arabic and Aristotelian sources. A striking feature was the idea that the two principles (sulphur and mercury) that made up metals were themselves composed from very small particles of elementary *minima* that in their turn were composed from the Aristotelian elements earth, air, fire, and water.

Geber, and likewise Sennert, pictured a hierarchical model of composition which Sennert was prepared to defend experimentally by appealing to the phenomenon of sublimation of the real substances sulphur and mercury. When heated in a still, these materials sublimed as sulphurous powder and discrete mercury droplets, so revealing their particulate nature. This clear resistance to further decomposition was evidence of their 'strong composition' which rendered them homogeneous. Fire (heat) allowed the lighter and more subtle particles to separate and ascend, leaving heavier, grosser particles behind. This implied that homogeneity depended upon size and weight.

Geber had also appealed to the calcination of metals as another form of sublimation. The experimenter detected a sulphurous smell as calcination proceeded, leaving a powdered and frequently coloured mass behind (the oxide), the different colours being explained as due to the different sizes and packing of the non-volatile sulphur left behind. The fact that metals could be regenerated unchanged after their solution in acids was additional clear evidence that they persisted through an alchymical experimental procedure.

Robert Boyle and the reform of chymistry

Our view of Boyle as 'the father of chemistry', and the most famous British natural philosopher before being eclipsed by Isaac Newton, has been completely transformed during the past thirty years. Michael Hunter and his acolytes have trawled through the huge archive of Boyle papers held at the Royal Society in London to produce a new edition of his published and unpublished works, correspondence, and laboratory notebooks. It turns out that Boyle was a great but complicated man who devoted most of his scientific life to drawing the attention of mathematically inclined natural philosophers to the merits of the chymists' activities for demonstrating a corpuscular theory of matter.

Boyle (1627–91), a man of insatiable curiosity, and a rather chaotic, literary output, was a Christian virtuoso who saw experimental philosophy as a way of understanding and justifying God's purpose and design. Besides being potentially useful, Boyle seriously believed that mechanical philosophy underwrote religion.

A rich and privileged Anglo-Irish aristocrat, he only began to be interested in science when he was in his early twenties. Before then he was a religious ascetic and cerebral moralist. His devoutness remained when he became an ingenious experimentalist devoted to the demonstration that an atomic or corpuscular philosophy was a better way of understanding God's creation than the pagan Aristotelianism and Paracelsianism of his day which, he believed, was propelling scholars towards atheism and infidelity.

Recent research has also thrown light on Boyle's education and the reasons he deliberately obscured his sources of information, so that he re-emerges as a less-original and even less-honest figure than he has been portrayed in traditional historiography. By portraying his views as novel (and, to be frank, less clearly and more verbosely than his sources), Boyle succeeded in beginning to build a wall between past and contemporary chymical practice and future chemistry.

Boyle's training as a chemist came from an American immigrant to England named George Starkey (1628–65), who wrote under the alchemical nom de plume of Eirenaeus Philalethes ('peaceful lover of truth'). It's a delicious irony that although Starkey taught Boyle chemistry, Boyle never knew that Eirenaeus was Starkey himself! Neither did Newton, whose chemistry, it turns out, was strongly influenced by Eirenaeus.

Much of the skill of chymists such as Starkey lay in their scholarly interpretation of other chymists' texts and the testing and modification of their procedures, as his surviving laboratory notebooks show. Understanding *Decknamen* was a hard job; hence Starkey's comment: 'God sells secrets for sweat'. The tragedy of Starkey's career was that in choosing to publish under a pseudonym he wrote himself out of history for over 300 years.

The very practical, lab-based nature of 16th-century alchemy and the reasonableness of attempts at gold making are now clear. (Compare the no-more fanciful hopes and aspirations of Victorian chemists that atmospheric nitrogen would one day be fixable for the manufacture of fertilizers.) Alchemists like Starkey and Helmont used the assayer's balance as an essential tool in mass balance reasoning concerning the relationship between reactants and products whatever transformations occurred in between to the human senses. That is, they were aware of a principle of conservation of mass. Alchemy, like the metallurgy that it drew upon, was quantitative; its practitioners tested their conjectures, and their theory-guided practice was founded on reproducible experiments.

For example, Helmont's obscure recipe for the preparation of the so-called 'alkahest' (matter degraded to its ultimate simplest form), which Starkey deciphered and revealed to Boyle, involved the preparation of an antimony–iron–mercury amalgam. Such an amalgam, following Starkey's recipe, when added to silver or gold does, indeed, lead to an apparent multiplication of these precious metals. But what Starkey and Boyle interpreted as *chryopoeia* (or metal making more generally) by the internal rearrangement of particles of the alkahest, we interpret as an efficient metallurgical process for the removal of silver and gold impurities from impure antimony ores.

Boyle's familiar demonstration that the volume of air trapped above a column of mercury was inversely proportional to the weight (pressure) of the mercury was also a good argument for the particulate nature of air—for how else could the air be compressed unless the particles of air were separated and not continuous? Similarly, his design of tests for acidity and basicity using coloured vegetable materials like litmus as indicators was to play an important role in the future analysis of materials.

Boyle used codes and ciphers to hide his alchemical interests and so bound himself to both a modern and ancient tradition. Although Boyle's rambling dialogue, *Sceptical Chymist* (1661), brilliantly criticized both the Aristotelian four-element theory and the Paracelsian system of three principles, he was clearly struck by Helmont's evidence that water might be the fundamental principle of matter, but he preferred to see this as evidence that there was a universal substratum of matter divided into corpuscles of different sizes and shapes. Analysis by fire (heating) did not produce elements or principles, but rearrangements of corpuscles making up matter.

Natural philosophers like Boyle, Descartes, and Newton were happy to think in terms of an imaginary world of atoms and corpuscles; but the practical chemists of the day, and particularly the iatrochemists and apothecaries, either had doubts or were indifferent to such speculations. Their sole aim was to produce useful and profitable medicines. It was the French pharmacists who drove the subject forwards at the beginning of the 18th century with their work at the Académie royale des science (founded in 1666) on the analysis of mineral waters and plants. Despite Boyle's scepticism, property-bearing principles continued to be useful to experimental chemists; but they were no longer 'simples' but categories or groups of stuffs. Salt, for example, embraced sea salt, saltpetre, vegetable and animal salts, and various alkalis.

In France in particular, chemistry came to acquire a public following that was reflected in the publication of large numbers of textbooks, many of which appeared in multiple editions. Chemistry there became important because its practitioners thought chemistry was worth teaching in order to encourage a greater public interest and to demonstrate its increasing economic importance. Chemistry had become an indispensable part of a Frenchman's liberal education.

The early laboratory

Despite the grand vision of Andreas Libavius in the 16th century, of a building with specialized room-laboratories, kitchen laboratory-workshops had remained standard for centuries and did not alter substantially until the 1840s. After all, cooking is a chemical process, and the problems of laboratory and kitchen maintenance are not dissimilar: drains, ventilation, furnaces for heating, raw materials, implements, and utensils. Not until the expansion of teaching (18th- and early 19th-century chemists like Berzelius commonly only took one pupil or apprentice at a time), and above all the introduction of organic chemistry, did laboratories need to be very large.

A small room with windows and/or a chimney for ventilation was quite adequate when apparatus was simple, consisting of flasks and alembics (distillation devices), and when chemical, pharmaceutical, and metallurgical operations were confined to heating, dissolving, and distillation. Not for nothing were chemists known as 'spagyrists' or workers by fire, or placed in the same category as smiths and farriers. Indeed, if we want to picture these early laboratories, we can do no better than to examine the interiors of the workshops of apothecaries and pharmacists, and not be misled by the often reproduced romanticized genre paintings of alchemists painted by Flemish artists like Daniel Teniers (1610–90).

Surviving prints of the Utrecht laboratory of Boerhaave's contemporary, Johann Barchusen (1610–1723), in 1698, or of the dispensing laboratory of the Amsterdam apothecary, Anthony d'Ailly, in the 1810s, both draw the eye's attention to the

6. The interior of a typical apothecary's laboratory and workshop which evolved into the modern chemical laboratory. The image is that of the Amsterdam apothecary Anthony d'Ailly in 1812. The workshop gave access his shop through the door on the left.

significance of furnaces and the ventilation of smoke and obnoxious fumes through large ventilation hoods (see Figure 6). The parallels with contemporary kitchens in the large houses of the period (e.g. that of the Prince Regent at the Royal Pavilion in Brighton) are striking. The French chef Marie-Antoine Carême (1784–1833) saw cooks as martyrs: 'It is the burning charcoal that kills us'. Indeed, the presence of deadly carbon monoxide in laboratories and kitchens from the use of charcoal as a heating agent made ventilation a priority.

Physics versus chemistry

Although Boyle had rejected the mathematical way as being too obscure for the general public, whereas the experimental was assimilable, many of his contemporaries found Boyle's work too hypothetical, and his rejection of analysis by fire and of the familiar elements and principles impracticable. If Boyle's support for a mechanical philosophy that resolved matter and its sensory qualities into shaped particles in motion seems like later physical chemistry that was not the way other practising chemists saw it. To them Boyle had thrown out the baby with the bath water.

The historian Victor Boantza has drawn attention in particular to the criticisms of an early member of the Académie de sciences in Paris, Samuel Du Clos (1598–1685). He was not at all impressed by Boyle's work and ideas, and refused to use any kind of mechanical philosophy in interpreting the programme of plant analyses he and others conducted in the Académie's laboratory. Du Clos preferred to interpret his analyses in terms of the Paracelsian *tria prima* that Boyle had so heavily criticized in his *Skeptical Chemist* (1661).

To Du Clos' way of thinking, Boyle's work showed ignorance of chymical literature and a lack of personal empirical practice. (This was perhaps a dig at Boyle's use of paid assistants to perform experiments under his direction.) Whereas his own work was empirical and demonstrable, Boyle's was abstract and ultimately merely descriptive.

As Boantza emphasized, both Boyle and Du Clos were in the business of reforming and modernizing chymistry: Boyle's strategy was to reduce chymistry to 'physico-chymical' discourse; Duclos' more practical aim was redefine traditional chymical philosophy, paying particular attention to elemental theories and modes of analysis. Both men agreed that neither the Aristotelian four elements nor the Paracelsian three principles were the ultimate constituents of compounds. But whereas Boyle rejected both theories completely, arguing that analysis by fire did not reduce bodies to their principles but more likely reduced them to their particles which then rearranged themselves, Duclos was prepared to use principles heuristically—like scaffolding—to help explain the results of analysis by fire (the dry way) or by distillation (the wet way). This operative view of principles was to lead to Lavoisier's pragmatic definition of elements in 1789.

French iatrochemists found the Paracelsian's solid salts of the greatest interest, and their composition of great significance. Pharmaceutical chemists at the Académie de sciences like Wilhelm Homberg (1653–1715) paid rigorous experimental attention to the identification, purification, and classification of salts and how they were derived from the neutralization between chemically active acids and bases, much of which was to provide the material for Lavoisier's *Traité élementaire de chimie* in 1789, where a salt was defined as a duality of acidic and basic oxides.

In view of Lavoisier's demand for precision in language it is interesting to note that the research that French pharmacists conducted in searching for drugs in plant extracts was logically self-contained and independent of any theory. The unsystematic nature of the terms used by these pharmaceutical chemists evidently did not prevent them from thinking consistently about the reciprocal reactions and compositions of the substances their terms designated.

Newton's chemistry and affinity

The reason why matter cohered and aggregated had always been rather a mystery to chymists. Aristotle's solution was simply that it was part and parcel of being 'earth' or 'water', and later Aristotelians appealed to water as a binding principle in mixed (compounded) bodies. However, cohesion was usually left as an occult power (in the sense that it was an unknown cause) and variously referred to as 'love/hate', or 'similitude/dissimilitude'. Following his successful mathematical analysis of planetary orbits in terms of a universal attractive gravitational force, Newton suggested that something similar existed between the particles of ordinary matter. In this case, however, the attractive forces extended only a short distance from the particles and varied in strength from one chemical species to another.

The investigation of chemical affinities became one of the absorbing problems of 18th-century chemistry. It seemed to French iatrochemists working with salts to be the basis for the relationships they uncovered. Étienne Geoffroy (1672–1731), who wrote of the *rapports* between chemical substances, produced the first table of affinities in 1718, and increasingly elaborate ones were produced from the 1750s onwards. Like the much later periodic tables, affinity tables effectively summarized the whole of chemists' knowledge concerning displacement reactions. The tables demonstrated the power of chemists to analyse the natural world into its components and how this information could be put to practical use—whether in preparing drugs or in explaining and improving chemical manufactures. On the other hand, the tables offered no theoretical insight concerning the causes of reactivity or comparative inertness.

Attempts by Newtonian chemists such as John Freind (1675–1728) and John Keill (1671–1721) to quantify chemical attraction soon proved hopeless. The Newtonian dream of a mathematical chemistry still lay two centuries away. On the other hand, the rich information gathered in affinity tables by such expert practical chemists as the Swede Torbern Bergman (1735–84) could also be recodified as instruction books on how to determine the composition of an unknown mineral or salt. One of the earliest and influential of these *Handbooks* of qualitative analysis (as they became entitled) was published in German by Heinrich Rose in 1829. As a result, the stinking poisonous gas, hydrogen sulphide, became the most familiar reagent used in chemical laboratories—especially after the Dutch pharmacist, P. J. Kipps (1808–64), developed an apparatus for its continuous production.

The appearance of systematic analytical tables that, like cookery books, enabled anyone to determine the composition of minerals, completed the emergence of a coherent chemistry of the mineral kingdom (eventually to be labelled inorganic chemistry). Principles of composition such as mercury, sulphur, and salt (or at least as rarefied or idealized versions of these real substances) expanded into a larger group of tangible homogeneous materials that were perceived corpuscularly and as the chymists' elements in explaining composition and the affinities of substances.

Chapter 3
Gases and atoms

Historians have often referred to a 'postponed revolution' in chemistry compared to the intellectual transformation of natural philosophy in the 17th century associated with Copernicus, Kepler, Galileo, Descartes, and Newton. The delay has been attributed to the fact that chemists had no understanding of the role of air in chemical changes until the mid-18th century.

While the centrality of the role of gases in promoting Lavoisier's new chemistry cannot be denied, historians have more recently also associated the late 18th-century transformation of chemistry with professional separations between medically oriented pharmaceutical chemists and more academic philosophical chemists. The latter ignored pharmacy and relegated it to an inferior intellectual position, while instead promoting the application of chemistry to the understanding and improvement of agriculture, mining, and technology generally.

The pneumatic revolution

Helmont had referred to the aerial products of reactions between acids and metals as *gas* because the products were chaotic and uncontrollable. They were first 'reigned in' and controlled by the Oxford-trained physician John Mayow (1641–79) in the 1660s.

Experiments with the new air pump devised by Robert Hooke and Robert Boyle in the early 1660s using birds, mice, and candles had led Boyle to conclude that atmospheric air acted as a transporting agent to remove impurities from the lungs to the external air. In his remarkable *Micrographia* (1665), Hooke outlined a theory of combustion that owed much to a contemporary meteorological theory that was based upon a gunpowder analogy. According to this 'nitro-aerial' theory, thunder and lightning were analogous to the explosions and flashing of gunpowder whose ingredients were sulphur and nitre. Since it was known that nitre lowered the temperature of water and fertilized crops, Hooke suggested that the nitrous particles trapped in the atmosphere were responsible for snow and hail and the vitality of crops. Such notions can be traced back to the Polish chymist Michael Sendovogius who, in 1604, had identified a *sal nitrum* (nitre) as a universal salt and component of the atmosphere.

It was Mayow who developed the nitro-aerial theory to its fullest extent in 1674, when he extended it to an even wider range of phenomena including respiration, the heat and flames of combustion, calcinations, deliquescence, the maintenance of body heat, the red colour of arterial blood, plant growth, and, once more, weather conditions. He recognized saltpetre as containing a base and an acid formed from one of air's constituents, the air itself being a medium formed from nitrous particles and materials that remained behind after respiration and combustion.

Mayow showed that when a candle burned in an inverted round-bottomed flask submerged in water, it consumed the nitrous portion of the air, causing the water level to rise. Combustion, he thought, involved the mechanical addition of the nitro-aerial particles to a metal which he knew from some of Boyle's experiments brought about an increase in weight. This explanation seemed to be confirmed by the fact that an identical substance (the calx) was produced when a metal was heated in air as when it was dissolved in nitric acid and then heated.

There is a superficial resemblance between Mayow's nitro-aerial theory and the century-later oxygen theory of combustion and respiration, but it is really only the transference properties that are similar. In any case, Mayow's model was mechanical rather than chemical. Nevertheless, Mayow was undoubtedly an ingenious experimentalist. Although he had not quite invented the pneumatic trough for collecting gases, his method of transferring an air (gas) from one glass vessel to another over water was original and influential. In particular, it stimulated the clergyman Stephen Hales (1677–1761) to devise a system of apparatus in which the aerial products of combustion could be 'washed' and so purified through water and collected in vessels for quantitative estimation (see Figure 6).

Inspired by Newton's mathematization of nature in the *Principia Mathematica* (1687), Hales was primarily interested in demonstrating that huge volumes of air were contained within otherwise solid minerals and plants—hence the title of his book on the subject, *Vegetable Staticks* (1727). Although he remained tied to the belief that the gaseous products he generated were pure air (that is, he failed to recognize that he had produced different gases), his 'pneumatic trough' proved to be the point of entry into distinguishing a range of gaseous products in the second half of the 18th century. For a time, too, his work seemed to reinvigorate the idea of the four Aristotelian elements. Like Helmont's willow tree experiment, Hales's work was based on underdetermined experiments (see Figure 7).

Experimentalists of Hales's era interpreted combustion in terms of a phlogiston theory that had been introduced to French and British natural philosophers from the work of a German chemist named Georg Stahl (1660–1734). Stahl's comprehensive theory was a development of a theory of composition that had been proposed by a German chymist and entrepreneur in 1669 (see Figure 8). Johann Becher (1635–82), contrary to Helmont, suggested that water was not the sole element and that minerals

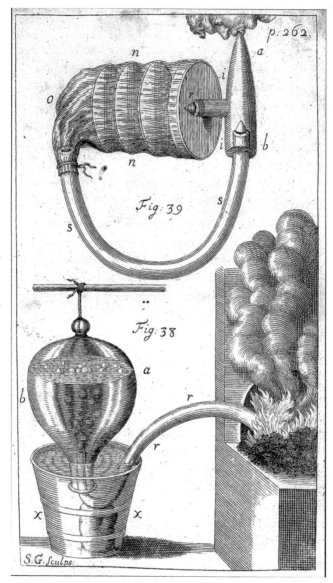

7. Apparatus for washing and collecting air fixed inside solid bodies.
From Stephen Hales, *Vegetable Staticks* (London, 1727).

**8. Georg Ernst Stahl, professor of medicine and chymistry at the
University of Halle. He developed the theory of combustion known as
the phlogiston theory in his edition of Johann Becher's *Physica
Subterranea* (Leipzig, 1703).**

were also generated from earth, of which there were three kinds.
One of these earths, which was fatty and sulphurous, he called
phlogiston. This earth was, he suggested, the cause of combustion
and combustibility.

Stahl developed this view further, and supposed that the
calcinations of metals were caused by the release of phlogiston
into the atmosphere from the heated metals from where it was
recycled into plants, animals, and minerals. The process of

calcination could be reversed by heating a calx (a simpler body than a metal) with another substance rich in phlogiston, such as charcoal derived from wood or bones. He listed many of the reversible transfers already familiar to chymists, but also noted that such reversals were not apparent in the vegetable and animal kingdoms where an organizing or vital principle was at work.

For Stahl, phlogiston explained not only combustion, but also metallicity, acidity, and alkalinity, the colours and scents of flowers, and chemical reactivity and composition. The theory was known to and adopted by French chemists in the 1730s, and from thence absorbed by British and Scandinavian pharmacists and chemists.

Combustion, composition, and language

The chemical revolution was not merely conceptual but also instrumental, in that it involved the practical ability to manipulate, weigh, and measure gases using accurate balances, glass apparatus, and eudiometers. The chemist who transformed our views of elements and composition, and reorganized the way that chemists communicated, was the French civil servant Antoine Lavoisier (1743–94).

For Lavoisier, as for his British contemporaries Henry Cavendish and Joseph Priestley, chemistry was purely a part-time avocation practised early in the mornings of days occupied with committees and working for a tax company. He owed much to a young wife who took notes, produced sketches of apparatus, prepared his papers for the press, and translated foreign works for him to read. In 1772 Lavoisier had noted the implausibility of a theory which suggested that something was lost (phlogiston) when a metal was burned in air when its weight actually increased rather than decreased. This was not a new observation, but this anomaly had been strikingly emphasized by the work of a Dijon lawyer.

In the spring of 1772, Lavoisier read an essay on phlogiston by Louis-Bernard Guyton de Morveau (1737–1816). In a brilliantly designed experimental investigation, Guyton showed that all his tested metals *increased* in weight when they were roasted in air; and since he still believed their combustibility was caused by the loss of phlogiston, he supposed that phlogiston was so light that it 'buoyed up' the bodies in which it was trapped.

Most members of the Académie des sciences, including Lavoisier, found Guyton's explanation absurd, and Lavoisier quickly deduced from his reading of Hales that a more likely explanation was that, somehow, air was being 'fixed' during the process of combustion and that this occluded air caused the increase in weight. It followed that 'fixed air' should be released when the calces of metals were decomposed, just as had been suggested by Hales's earlier experiments.

Lavoisier verified this in October 1772 by using a large burning mirror belonging to the Académie. When litharge (an oxide of lead) was roasted with charcoal a considerable volume of 'air' was, indeed, liberated. Lavoisier was still ignorant of the fact that his British contemporaries had shown that many different kinds of air were produced in chemical reactions.

In Scotland, a decade earlier, Joseph Black (1728–99) had succeeded in demonstrating that what we now call carbonates (e.g. magnesium carbonate) contained a fixed air (carbon dioxide) which was fundamentally different in physical and chemical properties from atmospheric air. Unlike the latter, it turned lime water milky and would not support combustion. A few years later, Henry Cavendish (1731–1810) studied the properties of a light inflammable air (hydrogen) which he prepared by adding dilute sulphuric acid to iron. These experiments were to stimulate the astonishing industry of the Unitarian clergyman, Joseph Priestley (1733–1804), who, between 1770 and 1800, prepared and differentiated some twenty new airs. These included (in our terminology) the oxides of sulphur and nitrogen, carbon monoxide, hydrogen chloride, and oxygen.

Hence, although largely unknown to Lavoisier in 1772, there was already considerable evidence that atmospheric air was a complex mixture, and that it would be by no means sufficient to claim that air alone was responsible for combustion. Lavoisier was soon made aware of his chemical ignorance, and with the firm intention of bringing about, in his own words, 'a revolution in physics and chemistry', he spent a year studying the history of chemistry—reading everything that chemists had ever said about airs since the mid-17th century and, significantly, repeating their experiments. Ironically, far from clarifying his ideas, his new familiarity with pneumatic chemistry led him to suppose that it was Black's fixed air (carbon dioxide) in the atmosphere that was responsible for the combustibility of metals and their increase in weight.

Two things forced Lavoisier to change his mind. First, he became aware that when heated the calx of mercury decomposed directly to the metal mercury without the use of the charcoal necessary with other metals. No fixed air was evolved. So how could the phlogiston theory be right? Here was a calx regenerating the metal without the aid of the phlogiston supposedly incorporated in charcoal.

The same phenomenon had come to the attention of Priestley. In August 1774 he heated the mercury calx in a closed vessel and collected a new air which he soon found supported combustion far better that ordinary air—hence he called it 'dephlogisticated air'. Priestley reported this observation directly to Lavoisier when he was on a visit to Paris in the autumn of 1774. Lavoisier's repetition of Priestley's experiment changed his mind.

In March 1775 Lavoisier announced to the Académie des sciences that 'the principle which combined with metals during calcinations and increased their weight' was 'pure air' and not any particular constituent of air. Clearly, Lavoisier was still confused and had to be corrected by Priestley later the same year, and with this stimulus Lavoisier was able to produce a new theory. He wrote in 1778:

> The principle which unites with metals during calcinations, which increases their weight and which is a constituent of the calx is: nothing else than the healthiest and purest part of the air, which after entering into combination with a metal, can be set free again; and emerge in an eminently respirable condition, more suited than atmospheric air to support ignition and combustion.

Because this 'eminently respirable air' burned carbon to form the weak acid, carbon dioxide, Lavoisier called the new gas 'oxygen', meaning 'acid former'. Thus by the late 1770s half of Lavoisier's chemical revolution was over. Oxygen gas was an element containing heat (or caloric, as Lavoisier called it) which kept it in a gaseous state. In other words Lavoisier redefined heat (or caloric) as something that altered the physical state of a substance without altering its chemical species, and not as something removed from bodies during combustion.

On reacting with metals and non-metals the heat was released and the oxygen element joined to the substance, causing it to increase in weight. In respiration, oxygen burned the carbon in foodstuffs to form carbon dioxide exhaled in breath, while the heat released was the source of animal warmth. Lavoisier and his fellow countryman, the mathematical physicist, Simon Laplace, demonstrated this quantitatively with a guinea pig in 1783 by measuring the amount of oxygen inspired by the pig and the amount of water its body heat produced when it was packed in ice—the origin of the term 'guinea pig'. Respiration, he concluded, was just a slow form of combustion. The non-respirable part of air (later called nitrogen, or azote in French) was exhaled unaltered.

On the face of it Lavoisier might seem to have only transferred the properties of phlogiston to oxygen gas, but there were major differences. Heat was absorbed or emitted during most chemical reactions and was present in all substances, whereas phlogiston was supposedly absent from non-combustible bodies. When added to a substance, heat caused expansion or a change of state, something not claimed by phlogistonists; above all, heat could be measured with a thermometer and oxygen with a eudiometer, whereas phlogiston could not.

The water problem

Lavoisier remained cautious and did not immediately suggest the abandonment of phlogiston. The main reason for this caution was that the phlogistonists could explain why an inflammable air (hydrogen) was evolved if a metal was dissolved in an acid whereas no air was produced when the calx (metal oxide) was treated with the same acid.

The problem, of course, was that moisture (a compound of hydrogen and oxygen) was so ubiquitous in chemical reactions that it was easy to overlook its production. It was Priestley who first noticed the presence of water when an electric spark was passed through a mixture of air and 'inflammable air' (hydrogen). In 1781 he mentioned this to Cavendish who repeated the experiment and reported in 1784: 'It appeared that when inflammable air and common air are exploded in a proper proportion, almost all the inflammable air, and near one fifth of the common air, lose their elasticity and are condensed into dew. It appears that this dew is plain water.' We are so used to knowing that water is composed from hydrogen and oxygen that it is hard to realize what an astounding discovery this was. How could two gases be transmuted into a liquid?

It was this same experiment which led the precise Cavendish to record that a bubble of uncondensed air remained, a fact reported later by Cavendish's biographer. The Scots chemist William Ramsay (1852–1916) was to recall reading this observation in 1894 when he showed that the bubble contained an unknown family of inert gases. For Lavoisier, Cavendish's work was evidence that water was not elementary. Assisted once again by Laplace, he showed that water could be synthesized by burning inflammable air and oxygen together in a closed vessel. For this reason, he renamed inflammable air, hydrogen, meaning 'water-former'. He could now explain why metals dissolved in acids to produce hydrogen whereas their calces did not. The hydrogen came not from the metal (as the phlogistonists supposed) but from the water in which an acid oxide was dissolved.

The new chemistry

Lavoisier was now in a position to bring about a transformation of chemistry by eradicating phlogiston from its vocabulary. Since all chemical phenomena were explicable without its assistance, it seemed highly improbable that the substance existed. He concluded a paper published in 1785 with a fine piece of rhetoric:

> Chemists have made phlogiston a vague principle which is not strictly defined and which consequently fits all the explanations demanded of it. Sometimes it has weight, sometimes it has not; sometimes it is free fire, sometimes it is fire combined with an earth; sometimes it passes through the pores of vessels, sometimes they are impenetrable to it. It explains at once causticity and non-causticity, transparency and opacity, colour and the absence of colours. It is a veritable Proteus that changes its form every instance.

By collaborating with a younger generation of assistants whom he gradually converted to his way of interpreting combustion, acidity, respiration, and other chemical phenomena, and by holding soirées where experiments and discussions could be held, Lavoisier won over a devoted group opposed to the idea of phlogiston. In 1788 Lavoisier and his companions founded a new quarterly periodical, *Annales de chimie*, to promote the new chemistry. Within a decade its contributors took the new chemistry for granted so that it became much more than a French journal promoting Lavoisier's views. Its successive editorial boards saw it as a European journal of international significance.

Three important converts were Guyton, Claude-Louis Berthollet (1748–1822), and Antoine Fourcroy (1755–1809). Guyton, in particular, was exercised by the inconsistent use of terms by chemists and pharmacists. Hitherto a substance might receive a different name according to the place where it was derived, or names might be based upon properties of smell, taste, colour, and usage. Following the inspiration of the Swedish naturalist Carl Linnaeus, who had begun to systematize the names of plants in 1737, Guyton suggested in 1782 that chemical language should be based upon three principles: substances should have one fixed name, names should reflect composition when known (or non-committal terms if unknown), and names should generally be chosen from Latin and Greek roots. It was while the four men were collaborating on a new language for chemistry that Guyton was converted to Lavoisier's new chemistry.

For his part, Lavoisier believed that language should express clear and distinct ideas. Together, in 1787, the four chemists published a 300-page manual of nomenclature reforms, one third of which consisted of a dictionary of new terms for old ones. For example, 'oil of vitriol' became sulphuric acid, and its salts sulphates instead of vitriols; 'flowers of zinc' became zinc oxide. The entire nomenclature was based upon the names of elements. Thus the elements oxygen and sulphur could combine to form either sulphurous or sulphuric acids depending upon the quantities of oxygen combined. These acids when combined with metallic oxides (the former calces) would form two groups of salts, the sulphites and sulphates.

Because the new language was also a vehicle for the non-phlogiston chemistry, it aroused much opposition. Nevertheless, through translation it rapidly became, and still remains, the international language of chemistry. Lavoisier's final piece of propaganda for the new chemistry was a textbook published in 1789 called (in English) *An Elementary Treatise of Chemistry*. Together with a much larger, less elementary text published by Fourcroy in 1801, this became a model for chemical instruction for several decades. In it Lavoisier defined the chemical element as any substance which could not be analysed further by chemical means. Such a definition enabled him to identify some thirty-three basic substances, including some which later proved to be compound bodies.

By the mid-1790s the opposition to phlogiston had triumphed, and only a few prominent chemists, such as Priestley (who emigrated to the United States in 1794) continued to believe in it. By then the French Revolution had put paid to the possibility that Lavoisier would apply his insights to fresh fields of chemistry. He was guillotined in 8 May 1794 for being a member of a firm of tax inspectors.

The atomic theory

John Dalton (1766–1844), the son of a Cumbrian Quaker handloom weaver, moved to Kendal in the heart of the English Lake District in 1781. There he became interested in meteorology, a subject that was to influence his thoughts on atomic theory. The records Dalton kept over a five-year period were published in *Meteorological Essays* in 1793. In the same year Dalton moved to Manchester as tutor in mathematics and natural philosophy at New College, where Joseph Priestley had taught in the 1760s. Dalton found Manchester so congenial that he decided to spend the rest of his life there. Not only was there an abundance of paid work in Manchester for private tutors because of the rise of an industrial middle class, but the presence of the Manchester

Literary and Philosophical Society, whose secretary Dalton became in 1800 and president from 1817 until his death, proved a convenient venue for publicizing his scientific work.

Like Priestley, Dalton is best described as a natural philosopher rather than as a chemist. Honed in the mechanistic tradition of Boyle and Newton, he believed that matter was particulate and endowed with powers of attraction and repulsion. However, Dalton moved away from the tradition of 18th-century matter theory which had emphasized the homogeneity of matter. Instead, he identified material particles with Lavoisier's elements, i.e. with substances that could not be chemically analysed. Thus, although Dalton spoke and thought of physical atoms, his chemical particles were much more akin to the *minima* or molecules of Aristotelian philosophers. Moreover, the identification of atoms and elements meant that Dalton, unlike some chemists such as William Prout (1785–1850), who speculated that Lavoisier's elements were polymers of hydrogen, accepted as many different kinds of atoms as there were chemical elements.

Dalton's theory probably originated from his meteorological study of air in the 1790s. Dalton wondered whether, if air consists of a number of gases as well as the water vapour precipitated as rain and dew, were these chemically combined or merely mixed together statically? If the latter, how is it that air is apparently homogeneous and does not consist of weighted layers of vapour, carbon dioxide, oxygen, nitrogen, and hydrogen?

Dalton's ingenious answer was to appeal to a model of self-repulsive air particles that had allowed Newton to deduce Boyle's law that pressure was inversely proportional to volume. In Dalton's revised model, however, the particles of each gas in the atmosphere were self-repulsive, though completely unaffected by each other. This produced mixing and atmospheric homogeneity. Although the model was later modified by ideas of diffusion and the kinetic theory of gases, chemists have always referred to the model as Dalton's law of partial pressures.

Since Lavoisier had explained the gaseous state of matter by the combination of heat (caloric) with elements or compounds—and heat particles were believed to be self-repulsive—Dalton's atmospheric model showed consistency. But in order to explain why self-repulsive forces differed from element to element and why, as his Mancunian friend William Henry pointed out, gases had different solubilities in water, Dalton was forced to conclude that the sizes of atoms of different elements varied. In order to calculate atomic sizes (volumes), densities and weights were required. Gas densities could be crudely determined by weighing equal volumes of gases on a sensitive balance, while atomic weights would be calculated from existing analyses if simple assumptions were made about atomic combination.

In about 1804 Dalton, having been through a long exercise to determine atomic sizes, realized that in calculating relative atomic weights he had produced a new quantitative basis for chemistry. With Dalton's permission, this was first publicized by Thomas Thomson in his *System of Chemistry* in 1807. Dalton explored the theory's implications in his own textbook in 1808 (see Figure 9).

Despite being plagued by methodological problems that were only resolved by the Italian chemist Stanislao Cannizzaro (1826–1910) in the 1850s, and by theoretical issues that were not resolved until the work of Rutherford and Soddy in the 20th century, Dalton's theory offered chemists a new, and enormously fruitful, model.

Dalton's originality lay in solving the problem of what philosophers have labelled 'transduction', i.e. deducing the existence of the unseen from macroscopic phenomena. The German chemist Benjamin Richter (1762–1807) had already demonstrated in 1792 that when acids neutralized bases to form salts, different quantities of acid were needed to neutralize a fixed amount of each base. The

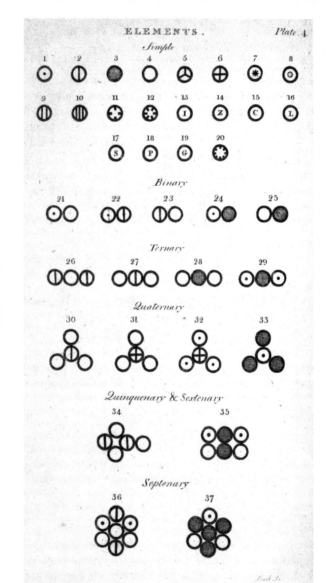

9. Dalton's atomic symbols and caption from *New System of Chemical Philosophy*, vol. 1 (London, 1808).

amounts were not arbitrary: a fixed (or *equivalent*) weight of acid reacted with a fixed weight of base.

This became known as a law of fixed proportions, and although Berthollet argued that the concentrations of the reactants ought to lead to variable proportions, the Spaniard Joseph Proust (1754–1826) provided sufficiently convincing evidence for fixed proportions as to convince the small chemical community of the time. (Berthollet's argument was nevertheless important, but it was another fifty years before it was developed into a law of mass action.)

Fixed proportions and equivalent weights implied, as Dalton saw, that matter was not continuous and that the ultimate particles of matter combined according to fixed rules. Knowing this, Dalton developed a way of calculating the relative weights of the ultimate chemical particles of matter from observations and measurements that were made in the laboratory.

To provide this calculus of chemical measurement Dalton had to make a number of simple assumptions about how atoms probably combined to form compound atoms, a process he labelled 'chemical synthesis'. In the simplest case, when only one combination of two elements could be obtained, he assumed it had to be a *binary* combination, unless there was reason to suppose otherwise. In other words, although two substances A and B might combine to form A_2B_2, it is simpler to assume that they will usually form AB.

In the case of water, for example, and bearing in mind that hydrogen peroxide was unrecognized before 1815, Lavoisier's analysis had shown that 87.4 parts by weight of oxygen combined with 12.6 parts of hydrogen. On the binary compound assumption, therefore, this ratio, H:O::12.6:87.4, must also be the ratio of the individual weights of the hydrogen and oxygen atoms that make up the water molecule. Since hydrogen was the lightest known substance, Dalton suggested that it should be taken as a standard of atom weights. Accordingly, if the atomic weight of hydrogen is taken as one unit, the relative atomic weight of oxygen is roughly seven. Improvements in analyses of water soon raised this to eight.

By drawing on the already extensive corpus of published analyses, Dalton was able to establish a lengthy list of relative atomic weights, which he first exhibited in Manchester between 1803 and 1804. Ironically, as a result of the emergence of electrolysis in the early 1800s, two British workers, Nicholson and Carlisle, had already demonstrated that two volumes of hydrogen and only one of oxygen were released when an electric current was passed through acidulated water. This might have suggested to Dalton, as it did to his Swedish contemporary, Berzelius, that water would be better expressed as H_2O. Dalton stuck to his simplest guns.

The French nomenclaturists of 1787 had noted various ways in which chemicals could be symbolized by geometrical patterns. However, although this represented an attempt to produce a simpler and more systematic notation than the alchemists, such symbols were inconvenient to reproduce in print and so never became firmly established. Typographical difficulties were also a factor in preventing Dalton's representation of elementary bodies by circles distinguished only by horizontal and vertical diameter lines, shading, or by placing an alphabetical letter within the circle. Although easy to draw and providing a concrete two-dimensional portrait of round atoms, they proved cumbersome to print (especially for multi-atom molecules) and never came into general use.

In 1813 Jöns Berzelius (1779–1848) introduced the modern notation whereby an element is symbolized by the initial letter of its Latin name, using two letters where elements might otherwise be confused: H, hydrogen; K, potassium (kalium); C, carbon; but Cu, copper (from cuprum). Berzelius' symbols, which could easily be arranged algebraically to represent compounds, became generally adopted from the mid-1830s onwards, when chemists also began to represent reactions by means of equations.

Are atoms split?

Dalton was well aware of the arbitrary nature of his rules of simplicity. In the second part of the *New System of Chemical Philosophy* published in 1810 he stated that water might be a ternary compound, in which case oxygen would have been sixteen times heavier than hydrogen: or, if two atoms of oxygen were combined with one of hydrogen, oxygen's atomic weight would be four. This uncertainty was to plague chemists for another fifty years.

Why, though, did Dalton not exploit Gay-Lussac's law of combining volumes, formulated in 1808? Dalton rejected the Frenchman's law on the grounds that if equal volumes of gases at the same temperature and pressure contain the same number of particles, n, then some of them would have to 'split', something that by definition no atom could do:

2 volumes hydrogen + 1 volume oxygen = 2 volumes water
$2n$ H particles + $1n$ O particles $\neq 2n$ HO particles

Thus the oxygen atom would have to divide in order to produce two water particles, occupying two volumes. Such division was made even more prohibitive by the fact that Dalton conceived atoms to be surrounded by atmospheres of heat which rendered them self-repulsive. Since there was no conceivable way two self-repulsive particles could he combined, the kind of atomic division apparently demanded by Gay-Lussac's law seemed impossible.

Dalton's views were reinforced by the electrochemical theory of combination introduced by Berzelius in 1813 (Chapter 4). Because like electrical charges were repulsive, atomic division was rendered implausible. Not until the time of Cannizzaro in 1858, was a way round this conundrum found in terms of molecular species. The trick was not to ask what held molecules together, but simply to accept, on chemical and physical evidence (kinetic theory), that most of the common elementary gases were diatomic.

Berzelius also introduced the notion of volume atoms. Assuming that the volume occupied by the atoms of different gases were all the same, because water consisted of two volumes of hydrogen and one of oxygen he arrived at H_2O as the formula of water; similarly he gave ammonia the formula NH_3 in contradiction to Dalton's formulae HO and NH. For this reason his atomic weights were radically different from those of Dalton's, even more so since he used an oxygen standard of sixteen (or sometimes one hundred) for relative weights rather than Dalton's hydrogen standard of one.

Berzelius devoted most of his life to determining accurate atomic weights, arguing that it was the most important object of chemical investigation and one worthy of 'unresting labour'. Despite his great accuracy, many of his atomic weights turned out to be incorrect. This was usually because he had to work from oxides. In the case of the two iron oxides, for example, he found that the quantity of oxygen (sixteen) that united with the same amount of iron were in the ratio of 2:3. From this he assumed the formulae were FeO_2 and FeO_3, rather than the FeO and Fe_2O_3 recognized later. Consequently Berzelius' atomic weight for iron was double the one agreed in later decades.

What was the chemical revolution?

Revolutions are exciting to write about, especially ones that are also set amid political upheaval; but in the case of the chemical revolution historians may have overlooked long-term trends and the fact that practical chemists had shunned the mathematical physicists' approach to material change.

They had seen Boyle's and Newton's approach as over-reductive in that it ignored the richly mysterious nature of chemical change. For them, in the words of French historian Sacha Tomic, chemistry was a practice involving 'operations, separations and combinations of substances that depended upon their properties and technical uses' rather than on the nature of matter itself. The American historian Victor Boantza sees what he calls 'the long chemical revolution' extending from Boyle to Lavoisier as a clash between physical and chemical perceptions. In that light, the phlogistonists' resistance to the Lavoisian approach can be seen as an attempt to hold a chemical as opposed to a physical method of investigation.

Despite the fact that the anti-phlogistonists won the debate, chemists' opposition to physicists' meddling with their science continued during the 19th and 20th centuries, and even continues today with chemists' refusal to allow the claim that chemistry can be reduced to physics. The chemical revolution can also be interpreted in terms of professionalization and the creation of new teaching institutions, career pathways, and journals. Nor can the role of the state be ignored, as governments (especially that of France) began to plan secondary and higher education and qualifying systems.

It is perfectly legitimate to argue therefore that Lavoisier's new chemistry was the apogee of an evolutionary, rather than revolutionary, process. The historiographical issue is best resolved by adopting the sophisticated approach of John G. McEvoy who has argued for a multivalent approach which gives equal weight, and interconnection, between theoretical, social, and institutional factors in the construction of modern chemistry.

Had he lived, the signs are that Lavoisier would have applied his new approach to vegetable and zoological chemistry. This was certainly the direction which many of his disciples followed and to which we can now turn.

Chapter 4
Types and hexagons

In recent years historians of chemistry have emphasized the changing roles between chemists and pharmacists during the chemical revolution, and the connection with the emergence of organic chemistry as a discipline in the 1830s.

It was Lavoisier's pharmacy-trained disciple Antoine Fourcroy who, as director-general of public instruction, did much to bring French science under state control. Fourcroy was the first chemist to establish that the bewildering variety of compounds that could be extracted from plants and animals generally contained just a few elements, namely carbon, hydrogen, oxygen, and nitrogen, and just occasionally sulphur and phosphorus.

Rather than talking of vegetable chemistry and zoological chemistry, Fourcroy divided chemistry into the two grand divisions of inorganic and organic. The latter term continued to refer to substances extracted from living entities until the 1830s. By then, however, chemists had prepared so many derivatives of extractable materials—that is, of substances that did not exist in the living world—that the term organic chemistry came to have a wider meaning. It was not, however, until 1858 that the German chemist Friedrich Kekulé (1829–96) defined organic chemistry as the chemistry of carbon compounds.

Chemists in the 19th-century pictured atoms, and used atomic and molecular models to create an academically and industrially significant discipline. 'Paper chemistry' and chemists' imagined worlds of molecules were to aid their studies and progress.

Liebig's achievement

The career of the Paris-trained German chemist Justus von Liebig (1803–73) encompassed innovation in teaching, important contributions to organic chemistry, and, above all, the application of chemistry to agriculture, physiology, medicine, nutrition, and industry, as well as to the popularization of chemistry.

Liebig's fame was not so much that he made any startling new chemical discoveries (see Figure 10). It was largely due to his

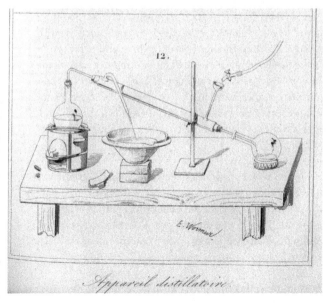

10. Water-cooled condenser usually attributed to Liebig because he used it extensively at Giessen.

demonstration with his Göttingen friend Friedrich Wöhler (1800–82) that it was possible to use the paper tools of Berzelian chemical symbols to make sense of analytical results by inspecting and juggling with the compositions of reactants and products.

Manipulating chemical symbols on paper was as significant as the introduction of written, as opposed to oral, examinations in universities during the same decades. As the German historian Ursula Klein has shown, the 1830s produced novel ways of individuating, identifying, and classifying organic compounds, chief of which was the exploitation of Berzelius' chemical formulae and their manipulation on paper in an attempt to understand the composition of the dazzling parade of new derivatives that were totally unknown in nature.

Liebig, Wöhler, and their French contemporary Jean-Baptiste Dumas (1800–84) excelled at this practice, and were considerably helped by the sophisticated method of organic analysis with the so-called *Kaliapparat* (potash bulb apparatus) that Liebig developed in 1830 when attempting to understand the composition of plant alkaloids. The replication of these gravimetric experiments in the 21st century has shown historians how accurate Liebig's results were for carbon, hydrogen, and oxygen content (though nitrogen content remained an acute problem).

Liebig and Wöhler first met in 1826, when they ironed out their difference of opinion over the apparently identical composition of Wöhler's silver cyanate and Liebig's preparation of silver fulminate. They agreed that the two silver salts were remarkable examples of different modes of combination among their elements carbon, hydrogen, oxygen, and nitrogen.

In 1830, Berzelius coined the word *isomerism* to describe the remarkable phenomenon whereby organic compounds with very different chemical and physical properties were composed from the same elements in identical proportions but in some unknown different physical arrangement. Wöhler produced another striking example in 1828 when he showed that the product of reacting silver cyanate with ammonium chloride was the 'organic' compound urea (and not ammonium cyanate as he had expected).

Later generations of chemist-historians saw Wöhler's 'synthesis' of an organic compound as the destruction of vitalism in chemistry, but it is now interpreted as a foundation myth invented by later chemists to provide an exciting take-off point. In fact, Wöhler's discovery of the isomerism between ammonium cyanate and urea did nothing to abolish the view that organic compounds extracted from vegetable and animals were under the control of something outside chemical knowledge, a vital creativity.

Instead, vital force simply withered on the vine through the development of synthetic organic chemistry. Isomerism was to be, and remains, a fundamental concept in understanding (and simplifying) the bewildering variety of the now millions of organic compounds that have been prepared by successive generations of chemists. Without this clarifying concept and the idea of structure that it eventually generated, organic chemistry would have long remained a 'dark forest', as Liebig and Wöhler aptly described it in 1832.

The development of a rapid and accurate method of gravimetric organic analysis using the *Kaliapparat* acted as a trigger for the explosion of organic (as opposed to inorganic) chemistry. The two techniques of paper chemistry and accurate compositional analysis forged the new discipline of carbon chemistry, as opposed to the tradition of vegetable and animal chemistry, and enabled chemists to classify and interpret analyses in terms of common groups or radicals and, later, in terms of 'chemical types'.

The 'Giessen model'

Liebig's contribution to the perfection of inorganic analysis and its dissemination must not be underestimated. At the University of Giessen, and building upon the long historical tradition of tests for mineralogical composition, he taught systematic methods of inorganic analysis, though he left it to pupils and assistants such as Carl Fresenius (1818–97) and Heinrich Will (1812–90) to publish these methods. (Fresenius went on to create a famous school of analytical chemistry in Wiesbaden that continues to this day.)

These systematic group separation methods (wet qualitative and quantitative analysis) were to be taught to every student of practical chemistry into the 1950s. The fame and celebrity status that Liebig sought as a young man came about through teaching these systematic methods of inorganic and organic analysis. Beginning with a majority of pharmacy students, he successfully attracted an international body of chemistry students to the University of Giessen, where he was appointed professor in 1824. From 1835 until he left for Munich in 1852, he engaged in line-production research investigating the chemistry of living systems of plants and animals. Whether Giessen was the model for future research schools has been the subject of great historical interest, since the 'Giessen model' rapidly spread far and wide.

Two French brigands

Charles Gerhardt (1816–56) had been trained and educated by his father to run a family white lead factory in Strasbourg, but Gerhardt fell so deeply in love with chemistry that he rebelled and took advanced studies with Liebig in Giessen and with Dumas in Paris. This led to a job in Montpellier, but disgusted with the poor facilities in such a provincial city he resigned in 1851, returned to Paris, and earned his living by private teaching in his own chemistry school. It was not until two years before his tragically early death that he was appointed a professor in his home town.

From 1844 onwards Gerhardt became a close friend of another rebellious chemist, Auguste Laurent (1808–53), who had trained with Dumas in Paris and worked in ceramic factories before teaching in Bordeaux. In 1846, appalled by the poor facilities at the University of Bordeaux, he joined Gerhardt in his school enterprise in Paris where they shared ideas about the reform of organic chemistry. Neither man respected Dumas, their former teacher, because he represented conservative chemical ideas and the establishment; both men were tactless and quarrelsome, but between them they transformed the way organic chemists viewed their chaotic subject.

By the 1850s, organic chemists were confident about the constitution and quantitative ratios of the elements composing an organic compound (though, in practice, their formulae might look different because there was no consistent system of atomic weights). By then chemists were also aware that elements 'clumped' together in regular patterns, and that there was an internal 'order' of elements within a compound, as indicated by the fact that the same empirical formulae had to be given to isomers that had distinctly different properties.

Berzelius had been a pioneer in differentiating the clumps within molecules by virtue of the dualistic system that, following Lavoisier's inorganic system, he promulgated from 1818 onwards. According to this electrochemical system of classification, organic compounds consisted of a positive hydrocarbon radical electrically linked to a negatively charged molecule that might contain more carbon and hydrogen together with oxygen, nitrogen, and other elements. This classificatory model was known as the radical theory; organic chemistry, in the words of Liebig, differed from inorganic chemistry because it involved compound, rather than simple, radicals. The radical model has persisted into the

21st century in the idea of 'groups', though these are no longer laboriously identified by 19th-century methods of analysis, but by their optical signatures that can be read spectroscopically or magnetically.

An alternative way of viewing the constitution of organic molecules had also matured from the 1830s. This had been generated from a discovery by Dumas in 1834 that electronegative elements or radicals such as chlorine, bromine, and cyanogens could be exchanged or substituted for hydrogen, in what Berzelius interpreted as the electropositive hydrocarbon portion of a molecule.

For Dumas, substances of the same chemical type clearly contained the same number of equivalents united in the same manner because they underwent the same fundamental reactions. For example, acetic acid, CH_3COOH could easily be transformed into CCl_3COOH, and both acids had very similar properties when reacting with other reagents. (Dumas actually wrote the two acids dualistically as $C_4O_4, C_4H_2H_6$ and $C_4O_4, C_4H_4Cl_6$, where $C = 6$.) Berzelius was initially not at all happy with Dumas' findings, but forced by the evidence, he began to elongate his formulae by allowing substitution in only the electropositive carbon grouping of an organic compound.

Meanwhile, Gerhardt, inspired by the ancient biological idea of a great chain of being, or ladder of nature, explored the idea of classifying compounds by a ladder of combination. Given the dazzlingly large numbers of organic compounds that chemists had already differentiated by the 1840s, Gerhardt supposed that a principle of plenitude operated, and that where two organic compounds differed by two units of carbon and hydrogen, another compound that differed by only one carbon and hydrogen unit must also exist. He soon called this the principle of homology.

Paraffins formed a series of compounds that differed by CH_2, and this applied equally to the organic acids formed from them. It was a simple idea, but one that had far-reaching implications for the learning and understanding of organic chemistry. It was the equivalent of the epiphany of the language student on learning that French verbs are only three in kind and that each conjugates in a regular way; or that there are only strong and weak verbs in German.

Substitution had already occurred to Dumas' pupil, Laurent. The latter opposed the dualistic views of Berzelius and, influenced by his studies of crystallography, where the principle of isomorphism revealed families of crystals with identical shapes despite their being different salts, Laurent chose to view organic molecules as unified or as 'single edifice' compounds, and not binary in composition.

He called these crystal-like edifices 'fundamental radicals'. Through substitution of elements within such unitary compounds, new compounds could be produced that he labelled 'derived radicals'. These derivatives, in their turn, played the same role as fundamental radicals, and in this way Laurent formed a natural classification of organic compounds. This system formed the substance of his posthumous treatise, *Chemical Method* (1854). When Laurent reviewed the warfare between the Berzelian dualists and his own advocacy of a unitary model he observed:

> Experiments went for nothing: dualism had sworn to uphold its position....I was an imposter, the worthy associate of a brigand [Gerhardt], and all this for an atom of chlorine gas put in the place of an atom of hydrogen, for the simple correction of a chemical formula!

The collaborators Gerhardt and Laurent realized their ideas of unitary compounds and homologous series were complementary, and that this offered a new way of classifying organic compounds.

Gerhardt first publicized the concept of homologous series in his widely read *Precis de chimie organique* in 1844, and he stressed its usefulness with respect to classifying and differentiating organic compounds of the form $[(CH^2) + H^2O]$, $[(CH^2)^2 + H^2O]$ and, more generally, $[(CH^2)^n + H^2O]$. When such compounds were oxidized, sulphonated, or halogenated, their empirical formulae retained the same form. This enabled Gerhardt to declare that if the composition, properties, and method of formation of a single substance obtained from one of the series of acids were known, it would be possible to predict the composition, properties, and mode of formation of all substances in a derived series.

Inevitably, Gerhardt's critics (and there were many) accused him of practising numerology rather than chemistry, but he was vindicated by his successful prediction of many compounds seemingly missing from the series of alcohols $[(CH)^2 + H^2O]$, or the prediction of the boiling points of compounds hitherto missing in a sequence. It should be emphasized that all this was empirical sorting and classifying, and the true significance of the repeating units of CH^2 only became apparent when Kekulé introduced the idea of carbon chains in 1858.

One of the reasons why the formulae of organic compounds printed in chemists' books and papers in the 1840s look so strange to us today is because chemists were obsessed with a formal analogy between inorganic and organic compounds while neglecting the assumption that equal volumes of matter in a gaseous or vaporous state contained the same number of molecules. Amadeo Avogadro's hypothesis which had been proposed in 1811 seemed physically impossible to Dalton and Berzelius, since it demanded that simple molecules like hydrogen and oxygen were binary, H_2 and O_2. However, if the atoms were positively or negatively charged, self-repulsion would drive them apart. When Berzelius and others calculated the molecular moieties of organic acids and their derivatives they chose to double the empirical formulae in order to ensure the presence of H_2O within the molecule.

Consequently, organic molecules were based upon a four-volume standard as opposed to the two-volume scale in which water was H_2O. Thus, sulphuric acid was ($SO_3 + H_2O$), whereas acetic acid's empirical formula was double that of today's, i.e. ($C_4H_6O_2 + H_2O$) rather than ($C_2H_2O + H_2O$). It was Gerhardt who drew attention to the anomaly that when four-volume formulae were used in organic chemistry, in reactions that involved the elimination of water (the moiety H_2O), it appeared as H_4O_2 (or $2H_2O$), and not as the H_2O it did in inorganic eliminations. A similar discrepancy appeared in organic reactions that involved the elimination of CO_2.

Gerhardt argued that consistency could only be achieved if *one* standard was used throughout chemistry, and that great simplification would be achieved by using two-volume formulae. Nevertheless, it took another twenty years before international agreement on a two-volume standard was reached. Ironically, Gerhardt himself never contemplated using Avogadro's hypothesis as a standard for determining molecular formulae.

One apparent difficulty with two-volume formulae was that organic monobasic acids (ones containing only one replaceable hydrogen atom) could not be written as if they were hydrated anhydrides. This was somewhat ironic, given that Humphry Davy (1778-1829)—the English discoverer of several post-Lavoisian elements like sodium, potassium, and iodine—had shown, despite Lavoisier, that the halogen acids (HCl, HI, etc.) did not contain oxygen. The difficulty was eventually resolved through the work of Thomas Graham and Liebig on bibasic and tribasic acids, which led to the notion that acids were better defined by their replaceable hydrogens. This reconceptualization, once again, undermined the simple dualistic electrochemical model of Berzelius and made the unitary model of Laurent much more acceptable.

Nevertheless, throughout the 1840s chemists such as the Lancastrian, Edward Frankland (1825–99), and the German, Hermann Kolbe (1818–84), continued to find the radical model useful. Both men thought they were able to prepare and isolate individual radicals of 'methyl', 'ethyl', and so on, just as if they were new simple elements. Kolbe did this by the electrolysis of organic acids; Frankland by reacting the iodide salts of organic acids with metals such as zinc in sealed tubes under pressure:

$C_4H_5I + Zn \rightarrow C_4H_5 + ZnI$
ethyl iodide + zinc → 'ethyl' + zinc iodide

It soon transpired that the boiling points of these hypothetical radicals did not fit in with the notion of homologous series, and that what Frankland and Kolbe had really prepared were the stable dimers of the supposed methyl and ethyl radicals: i.e. not C_2H_3 and C_4H_5, but C_2H_6 and C_4H_{10}, ethane and butane. An unintended consequence of this attempt to isolate radicals was Frankland's discovery of a range of organometallic compounds (his term) that were to prove of great interest in the 20th century.

Chemical types

By the 1850s, Gerhardt had begun to suggest a new, far-reaching way of classifying organic compounds that he named the type theory. Influenced greatly by Laurent's unitary idea of molecules, in which multiple substitutions of different elements and radicals were possible, and by the previous work of Dumas, he suggested that all organic molecules could be modelled on relatively few inorganic prototypes; organic molecules were simply substitution derivatives of the inorganic molecules hydrogen (H_2), hydrogen chloride (HCl), water (H_2O), and, later, ammonia (NH_3).

11. Alexander Williamson (1824–1904), professor of chemistry at University College London.

Why four types when two might have done? The answer lies in the fact that although Gerhardt and Laurent had undermined the simple form of Berzelius' electrochemical theory, they realized that molecules did display polarity. Thus H_2 was the type that displayed the least polarity, and that when the hydrogens were replaced by hydrocarbon radicals the relatively inert paraffin and olefin homologous series were produced. Substitutions in HCl produced the more active and polar organic salts such as ethyl iodide, C_2H_5I; and H_2O, as Gerhardt's British disciple Alexander Williamson (1824–1904) demonstrated in 1850, provided an extremely elegant way of explaining the makeup of the organic alcohols and ethers (see Figure 11). Types literally obliged inventiveness among the typographers in the printing works, as elaborate brackets began to pepper chemists' printed papers:

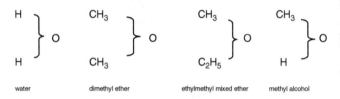

water   dimethyl ether   ethylmethyl mixed ether   methyl alcohol

Williamson stated proudly:

> The method here employed of stating the rational constitution of bodies by comparison with water, seems to me to be susceptible of great extension; and I have no hesitation in saying that its introduction will be of service in simplifying our ideas, by establishing a uniform standard of comparison by which bodies may be judged of.

If acetic acid is written as

$$\left.\begin{array}{l} C_2H_3 \\ \\ H \end{array}\right\} O$$

there should be an anhydrous form (acetic anhydride)

$$\left.\begin{array}{l} C_2H_3O \\ \\ C_2H_3O \end{array}\right\} O.$$

It was Williamson's particular triumph to predict this anhydride's existence in 1851, and it was duly prepared by Gerhardt in the following year. Williamson's work on ethers was soon followed by Hofmann's work at the Royal College of Chemistry in London on the nitrogen-containing amines. Here ammonia, NH_3, became the model type (where R, R', R" stand for various radicals):

$$
\left.\begin{array}{l} H \\ H \\ H \end{array}\right\} N \quad \left.\begin{array}{l} CH_3 \\ C_2H_5 \\ C_3H_7 \end{array}\right\} N \quad \left.\begin{array}{l} R \\ R' \\ R'' \end{array}\right\} N
$$

Gerhardt lived just long enough to learn that Williamson's colleague, William Odling (1829–1921), proposed adding a marsh gas (methane) type, but not long enough to witness how Kekulé used this to deduce the idea of carbon chains in 1858. The latter signalled a further triumph for Gerhardt's reforms.

By hypothetically substituting radicals such as ethyl into water, he achieved a rapprochement between the hitherto opposed radical and type classifications. Moreover, by subsuming organic compounds under four inorganic types Kekulé effectively brought about a unification of chemists and of chemistry. Organic chemistry, in Kekulé's hands, was to be the chemistry of carbon compounds—as he subtitled his textbook, *Lehrbuch der Organischen Chemie* in 1861.

Frankland's introduction of valency

Type theory (which soon embraced so-called 'mixed types' in which, for example, the water and ammonia types were bolted together) implicitly contained within it the important concept of valence, or the combining power of individual atoms.

Frankland, whose fame derived from his exploration of organometallic compounds like tin diethyl and stanethylium, had been the first to address this in 1852. He pointed out that the elements within an inorganic or organic compound appeared to have a definite or fixed affinity for other atoms, or 'only room, so to speak, for the attachment of a fixed and definite number of the atoms of other elements'. Thus hydrogen seemed to fix only *one* other attachment, oxygen *two*, and nitrogen as many as *five*.

Frankland called this regularity 'atomicity', but it was soon being called 'valence' or 'valency', from the fact that the attachments could be regarded as one, two, or more units that were 'equivalences' of hydrogen. Kekulé, following Odling, took this further by suggesting that carbon had the ability to unite with four equivalents of hydrogen and combine with itself to form chains such that two joined carbons still had a combining capacity or valence of six, and a chain of three carbons, eight, and generally $nC = 2n + 2$. Where this did not work, namely with the olefins, C_2H_4, and acetylenes, C_2H_2, Kekulé boldly introduced the idea that in such compounds two carbons could also mutually satisfy two or three valences and so form double or triple bonds. (The term 'bond' had been introduced by Frankland in 1866.)

Although the development of valence as an architectural concept for linking atoms together within a molecule owed more to Kekulé than to Frankland, it was the teaching and examining position of the latter as Hofmann's successor at the Royal College of Chemistry that enabled him to spread the concept of valency through textbooks and government examinations.

Kekulé's news from Ghent

Friedrich Kekulé had intended initially to train as an architect, but studied chemistry with Liebig at Giessen instead. Liebig sent him to London for further experience, and there he came under the influence of Odling and Williamson and their enthusiastic support for the unitary and type theories of Laurent and Gerhardt. After further private studies in Heidelberg, he was appointed professor of chemistry at the University of Ghent in

French-speaking Belgium in 1858. He finally returned to Germany as a professor in Bonn in 1867.

Kekulé's theoretical insights into the classification of organic compounds, valence, and structural chemistry, and his suggestion of the hexagonal formula for benzene, transformed organic chemistry. In his great paper of 1858, 'On the constitution and metamorphoses of carbon compounds and the chemical nature of carbon', in which he extended the quadrivalence of carbon in methane to all of its compounds, Kekulé stressed how indebted he was to the English and French schools of chemistry for his interpretation. With hindsight it can be seen that at one stroke organic chemistry had been unified: chemists no longer needed to separate 'types' for paraffins, ethers, amines, etc.—all organic compounds were now embraced within the idea of carbon chains (catenation) and the notion of carbon's tetravalence.

In an after-dinner speech in 1890, Kekulé's recalled daydreaming on an omnibus while he was a postdoctoral student in London in 1855. In his imagination he had visualized carbon atoms linking together to form long chains. Structural organic chemistry demands imagination and the ability to think architecturally, but Kekulé's ideas did not involve connecting carbon atoms by lines in the 1850s. Instead, he imagined segmented wormlike entities made up from sub-atoms. The sausage-shaped graphic formulae (Figure 12) that he first used in lectures in Heidelberg in 1857 followed from this model, and seem confirmed by Kekulé's statement in 1867 that 'polyvalent atoms, with respect to their chemical value [valence], can be viewed in a sense as a conglomeration of several mono valent atoms'. The graphic formulae that Kekulé used in his *Lehrbuch der Organischen Chemie* were printed visualizations of this model, and he continued to use type formulae as a means of classification.

Explicit line-connected chains were the independent idea of a Scots chemist named Archibald Scott Couper (1831–92) who was

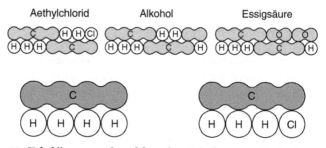

Aethylchlorid   Alkohol   Essigsäure

12. **Kekulé's sausage-shaped formulae. Univalent atoms like hydrogen were represented by single circles, and multivalent atoms like carbon by overlapping circles.**

working in Adolphe Wurtz's laboratory in Paris between 1856 and 1858. Couper's paper, boldly titled 'On a new chemical theory', had its publication unfortunately deferred until after Kekulé's own paper had been published. Mental illness then blighted Couper's blossoming career, and his significant contributions to paper chemistry remained virtually unknown until the 20th century. Nevertheless, his graphic notation was noted by the professor of chemistry at Edinburgh, Alexander Crum Brown (1838–1922), who enthusiastically deployed such formulae in his thesis in 1861. These line-drawn graphic formulae were then adopted by Frankland in his lectures at the Royal College of Chemistry in London, as well as in his research papers. Frankland was also an influential examiner for state tests, and in this way he ensured the rapid adoption of such graphic formulae in the 1860s.

The benzene ring and 'Kekulé's dreams'

Historians of chemistry have tended to stress the slow evolutionary continuity of Kekulé's work in organic chemistry rather than the sudden emergence of particular insights. We cannot be certain exactly when Kekulé hit upon the structure of benzene as a closed chain of six carbon atoms, though it probably occurred to him around the time of his first marriage in 1862.

Towards the end of his life, in typical self-deprecatory fashion, he said that, while dozing, he had imagined a chain of dancing carbon atoms forming a closed circle, like a snake eating its own tail. Historians have long debated the validity and nature of Kekulé's two 'dreams' on historical and chemical grounds. It must also be recognized that the 'Benzolfest' of 1890 at which Kekulé recounted his daydreams was designed less to honour Kekulé than to impress participants and newspaper readers of the central economic significance of chemistry for the Reich. Consequently, accounts of the speeches have to be used with caution as historical texts.

Despite his declining powers, lethargy, and fixation with family history and ennoblement—which his radical theorist rival Kolbe exploited in inexorable criticism of Kekulé's structure theory—Kekulé did have some significant achievements after he moved to Bonn in 1867. In 1872, he proposed a daring dynamic oscillation formulation for the structure of benzene that explained the embarrassing lack of isomeric disubstituted derivatives in benzene that seemed otherwise possible.

In the early 1880s, when several alternative possibilities had been touted for benzene's structure, including a prism formula proposed by Ladenburg, Kekulé demonstrated a return of his old powers when he showed that a series of experimental transformations of pyrocatechol and quinine were best and most simply explained if benzene was a hexagonal closed carbon chain (Figure 13).

While it is convenient to suppose that it was Kekulé's architectural training that helped him conceive molecular structure and to play with molecular models, what is more striking is the view he acquired from Williamson of the dynamic nature of molecules. Architecture is essentially static, whereas Kekulé's conception of structure was much more fluid and imprecise. To that end, the visionary giddy molecular dances cited in his speech at the Benzolfest in 1890 ring true.

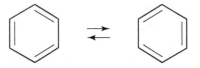

13. Kekulé's oscillating benzene rings. Collisions between the carbon and hydrogen atoms in the hexagonal ring were supposed to lead to fluctuating positions of the double bonds. This 'resonance' accounted for the absence of isomeric disubstituted benzene compounds in the first and second positions of the apex.

The triumph of structural theory

One of the strongest supporters of a structural view of organic chemistry was the German-trained Russian chemist, Alexander Butlerov (1828–86). It was he who pointed out how the theory would explain the existence of isomers. Two substances with an identical empirical formula like ethyl alcohol and dimethyl ether, C_2H_6O, had two different structures, CH_3CH_2–OH and CH_3-O-CH_3, and it was the different positions of the atoms within the molecules that gave rise to their different properties.

Further confirmation of structuralism came with the elucidation of stereoisomerism, the phenomenon whereby two isomers differed only in one physical property, namely the way their crystals rotated the plane of polarization to the left or right, as Louis Pasteur (1822–95) first elucidated in 1848. In 1874, the Dutch organic and physical chemist Jacobus van 't Hoff (1852–1911) explained Pasteur's observations by suggesting that the four valence bonds of carbon were directed towards the apices of a tetrahedron. This would allow the possibility of asymmetry and the existence of two configurations of a molecule that were the mirror images of each other. Soon chemists were applying these geometrical ideas to other atoms like nitrogen and sulphur, and stereochemical research became extremely fashionable until it culminated in the coordination theory of molecular structure developed by the Swiss chemist Alfred Werner (1866–1919) from 1891 onwards.

In Werner's theory, which applied particularly to the kinds of organo-metal compounds investigated by Frankland, molecular groups (later termed ligands) could be distributed around a central metal in various geometrical forms. Geometry among chemists became quite fashionable. The Germans Viktor Meyer (1848–97) and Adolf von Baeyer (1835–1917), for example, introduced the idea of steric hindrance to explain how ringed structures might set themselves in one particular rigid form because a large molecular group prevented the rotation of the two parts of the molecule into a different (isomeric) structure; or at other times be quite unstable because the bonds between the atoms were under strain.

Probably the greatest triumph of structuralism was to be seen in the work on sugars of Emil Fischer (1852–1919) in the 1880s. Fischer found that there were sixteen isomers of the sugars that shared the empirical formula $C_6H_{12}O_6$; the only difference between them was the extent of their optical activity produced by the fact that each sugar had four asymmetric carbon atoms.

Bringing chemistry to order

In a short account of the development of chemistry it is impossible to do justice to the hundreds of minor chemists without whose contributions the edifice would never have been completed. To take but one example: Leopold Gmelin (1788–1853) was a member of an important dynasty of physicians and apothecaries going back to the 16th century and Wöhler's teacher at the University of Heidelberg. He was described by a contemporary as 'one of chemistry's athletes'. He was certainly an important teacher: his pupils included Wöhler, Will, Bunsen, and Williamson, as well as several generations of influential pharmacists. As Berzelius famously remarked: 'anyone who has studied chemistry with Gmelin will have little to learn from me'.

Today, Gmelin is chiefly remembered by historians of chemistry for three reasons: his skilled analytical work on the chemistry of digestion which he undertook with his physiological colleague Friedrich Tiedemann in the 1820s; his remarkable ordering of the elements into chemical triads that foreshadowed the later periodic system of classification; and for his monumental *Handbuch der Chemie* which went through four editions during the author's lifetime and continues to be produced by the Gmelin Institute in Berlin to this day.

Gmelin's original idea was to summarize all known chemical knowledge, but he soon found it was impossible for one man, however athletic, to keep up with the phenomenal progress of organic chemistry in the 1830s. The last edition of the *Handbook*, which was translated into English by Henry Watts (librarian of the London Chemical Society) between 1848 and 1872, was largely restricted to inorganic chemistry. Gmelin's decision gave the Russian-German organic chemist Friedrich Beilstein (1838–1906) the opportunity to devise a similar encyclopaedic survey of organic chemistry from 1881 onwards. Beilstein's plan was to provide an account of the physical and chemical properties of every organic compound described in chemical literature and identified by its molecular formula. Before his death Beilstein handed the enormous task of compilation over to the German Chemical Society that Hofmann had founded in 1867. 'Beilstein', as it is called by chemists, continues to this day in online searchable form.

An acute problem faced by compilers of such handbooks, let alone laboratory workers, was agreement on the naming of compounds for indexing purposes. In 1892 an international group of organic chemists and editors of chemical journals met in Geneva to forge agreement on nomenclature. They agreed to use the structural theory as the basis for giving systematic names to chemicals, the idea being that the name itself would enable any chemist to deduce a compound's structural formula and to find it in Beilstein.

Common diethylether, for example, became ethoxyethane and common names such as olefiant gas and acetylene became ethane and ethine respectively. The citric acid of lemon juice, $C_6H_8O_7$, became the more formidable 2-hydroxypropane-1,2,3-tricarboxylic acid. The Geneva conference was necessary, but chemists paid a heavy price for inventing systematic names for unambiguous structures—namely the loss of public understanding. Doubtless this explains the persistence of a common nomenclature. Water remains water and only rarely is it necessary to write dihydrogen oxide to avoid ambiguity.

During a 19th century of chemical 'giants' it is worth emphasizing Williamson's pivotal role in the development of 19th-century chemistry. Not only did his demonstration of the power of the water type help lead to the concepts of valence and structure, but the elegant work on etherification also suggested how chemists might in future investigate the mechanism of chemical changes as individual atoms exchanged positions in stages through intermediate compounds.

His visualization of atoms in motion, and reactants and products in dynamic equilibrium, were to influence the development of physical chemistry profoundly. As a firm believer in the existence of atoms, Williamson did much to eradicate the predominant scepticism of his fellow chemists. These profound academic contributions were accompanied by a remarkable personal generosity of spirit. He personally educated successive groups of Japanese students at University College London, many of whom kick-started the modernization of hitherto isolated Japan on their return. Their descendants ensured Japan's place at the forefront of later chemistry.

Chapter 5
Reactivity

There has always been a close relationship between the teaching of chemistry and physics. Indeed, since the 18th century what are now seen as separate experimental disciplines were taught in a unified way. Robert Bunsen was famous for saying that a chemist who professed not to know any physics could not be a genuine chemist.

The rise of organic chemistry soon put a strain on the amount of information and experimental demonstrations that a single professor could lecture on. It was this fact, together with the limited opportunities for chemists trained in Germany to find academic posts, that caused several chemists in the 1850s to transform themselves into physicists.

In teaching the elements of light, heat, sound, electricity, and magnetism they frequently became interested in the physical properties of matter and how these affected chemical transformations. The result was the emergence of a third specialized area of chemistry known as physical (or, sometimes theoretical or general) chemistry. Physical chemistry, not inorganic or organic, was destined to become the basis upon which chemistry was to be taught. These developments necessarily led to changes in the design of chemists' work spaces for teaching and research.

Why do homogeneous substances interact? By the mid-19th century many different kinds of chemical change had been recognized and systematized in textbooks. Sometimes elements combined together directly; at other times one element substituted for another. Compounds could dissociate and then reassociate, or two compounds could undergo a double decomposition with an exchange of partners. Elements and compounds could be oxidized or reduced by the addition of oxygen or the removal of hydrogen; and sometimes a compound could be induced to undergo isomeric change.

It became the task of physical chemists to explain these different transformations in terms of exchanges between atoms and molecules powered by energy changes and the shifts in equilibrium that underlay all reactivity. Physical chemists found ways of expressing chemical change in mathematical terms, and so brought generalization and systematization to chemical practice. They entered an ideal, imagined world that provided a guide to the behaviour of real substances.

Berthollet's investigations of reactions at the beginning of the 19th century had led him to the idea that chemical transformation depended upon physical conditions and that affinities could be satisfied in dozens of different ways, leading to the claim that compounds were not formed in definite proportions. Such a view seemed impossibly complicated and was quickly replaced by the Daltonian view that chemical changes proceeded in definite quantities or proportions.

Consequently, it was not until mid-century that a few chemists returned to the question first raised by Berthollet: what were the conditions for chemical equilibria, and what were the mechanisms involved in chemical reactions when one homogeneous stuff interacted with another to produce other completely different homogeneous substances?

Are physical properties regular?

One early physical research programme was established by Hermann Kopp (1817–92), a colleague of Liebig's at Giessen who subsequently joined Robert Bunsen (1811–99) at Heidelberg. Kopp was interested in how physical properties like melting and boiling points were affected by changing composition. Could it be shown, he asked himself, that the physical constants associated with substances, like melting and boiling points, were functions of the chemical nature of molecules?

It seemed for example that in the homologous series identified by Gerhardt, melting points (a useful way of establishing the purity and identity of an organic compound) increased as one paraffin was transformed into another by the addition of CH_2. As Kopp demonstrated in the 1840s, a difference of nCH_2 corresponded to about $n19°C$ in boiling point for organic paraffins, acids, esters; and that a difference of n carbon atoms produced boiling point differences of $n29°C$; or that the additions of hydrogen atoms alone to a compound caused compounds to boil at $n5°C$ lower. Physical properties were clearly a function of a compound's stoichiometry.

Kopp spent nearly all his career when he was not investigating the history of chemistry (in which he was also a pioneer) making meticulous experimental determinations of physical properties with the aim of being able to offer arithmetical or geometrical series formulae to calculate the physical properties of newly discovered compounds. It is a curious fact that, at the same time as Kopp's programmes, other chemists, inspired by Prout's hypothesis, were trying to formulate arithmetical relationships between the atomic weights of elements. A good deal of what was then referred to as 'chemical physics' involved Pythagorean musings.

Kopp died in 1892 without having produced a satisfactory and exact way of correlating physical properties with molecular composition;

but he produced a wealth of data that proved extremely useful for analytical and synthetic chemistry. In the 20th century his work stimulated the production of compilations of physical data such as Kaye and Laby's *Tables of Physical and Chemical Constants* in 1911; this has remained in print ever since, and expanded from 150 pages to over 600 pages in successive editions.

The broad highway of thermodynamics

The most important area of chemistry to be influenced by physics was thermodynamics, a science that arose when physicists and engineers began to concern themselves with the efficiency of the steam engines and electric motors that had come to dominate industrialized parts of the world.

Chemical thermodynamics had begun earlier with thermochemistry. This can be traced back to the 18th-century studies of specific and latent heats made by Joseph Black, who showed that different substances had different capacities for absorbing heat. In addition, large amounts of heat that were not registered by a thermometer were required to melt a solid or boil a liquid. Specific heats proved useful when the École Polytechnique-educated Pierre Dulong (1785–1838) and Alexis Petit (1791–1820) found a simple relationship between an element's specific heat and its atomic (or equivalent) weight. Their empirical relationship, atomic weight multiplied by specific heat equals 6.4, proved useful in determining atomic weights, as Berzelius showed.

As the word suggests, thermodynamics concerns the transformation of heat into the useful work of a machine. By the 1840s, physicists like Rudolf Clausius (1822–88) in Germany had established the first law of thermodynamics: that energy (a measure of the ability of molecules in motion to generate heat and power) was conserved. He meant by this that whenever work was done through the agency of heat, a quantity of heat was consumed that was proportional to the amount of work done; conversely, the expenditure of an equivalent quantity of work produced an equal quantity of heat.

Clausius promoted a new, molecular way of understanding heat as the mechanical motion of particles, as opposed to Lavoisier's imponderable caloric. Clausius knew sufficient chemistry to perceive chemical change as dissociation, and was encouraged by Williamson's work in England on the mechanism of etherification. Heat drove molecules apart and overcame the intrinsic affinities or attractions between coupled particles. There came a stage in a reaction, however, when the forces of attraction and thermal dissipation entered equilibrium.

In 1873 another German chemist, August Horstmann (1842–1929), took this much further by asking himself what the extent of dissociation was in any particular reaction and equilibrium. Noting the work of Clausius, he posited that the limit of any dissociation depended upon the entropy of the reaction reaching a maximum.

Although Lavoisier's purpose in using a guinea pig was to compare its body heat with the heat evolved when carbon was burned to carbon dioxide (see Chapter 3), the experiment led many other chemists to investigate the heats of reactions. Between 1838 and 1840, Berzelius' Russian pupil, Germain Hess (1802–50), showed that the heat evolved in a chemical reaction (later named *enthalpy*) is always constant no matter whether the reaction takes place in one or several steps. This soon became known as Hess's law, one of the many eponymous rules that are ubiquitous in chemistry. With the physicists' enunciation of the first law of thermodynamics it became clear that Hess's law was simply a consequence of the fact that energy could neither be created nor destroyed.

Chemists were far slower to adopt the second law of thermodynamics, mainly because they could not understand the concept of entropy or see how it could be useful to their laboratory practice. This was true even of those chemists who had the necessary mathematical knowledge to follow thermodynamic notation and equations. This was especially the case with the remarkable rigorous mathematical treatment of chemical equilibria made by the American physicist J. Willard Gibbs (1839–1903) in the mid-1870s. His three papers laden with 700 equations appeared baffling to ordinary chemists, and even Clerk Maxwell, Britain's greatest physicist since Newton, found them incomprehensible. However, once the Austrian physicist Ludwig Boltzmann (1844–1906) had described the motion of gas particles as statistical averages in the 1870s (statistical mechanics), it became easier to interpret entropy in terms of disorder. Thus, the higher the entropy, the higher the disorder within a molecular system: entropy, then, was the driving power of chemical reactions. In the physical changes from solid through liquid to gas, entropy increased as the molecules became increasingly disordered.

Textbooks of thermodynamics suitable for chemists only began to appear in the 1900s as chemistry departments began to teach mathematics for chemists. However, notations, symbols, and positive and negative signs were used inconsistently, and it was not until the American chemists Gilbert N. Lewis (1875–1946) and Merle Randall (1888–1950) published their text *Thermodynamics* in 1923 that consistency was established.

One of their innovations that gave clarity to physical chemistry was to emphasize the idea of the 'free energy' of chemical substances, that is, the maximum of work that can be drawn from any chemical process. Free energy was not a new idea, having been introduced by Gibbs and others in the 1870s, but it was Lewis and Randall who demonstrated how it could be used to produce a deeper understanding of chemical reactions and produce useful numerical data.

In 1909, following suggestions from German mechanical engineers, the Dutch physicist Kamerlingh Onnes (1853–1926), who is best known for his liquefaction of helium in 1908, suggested that the portion of free energy that could be produced in actual chemical reactions, namely the total heat of a substance available between agreed temperature points, should be named the *enthalpy* (Greek, 'heat inside') of a substance. The term was not much used in chemical literature before the 1950s, by which time the historical ideas of chemical affinity had become obsolete and replaced by thermodynamics.

From the 1960s onwards, thermodynamics, and the calculations and data it generates, have been presented in terms of 'the three "e"s': energy, entropy, and enthalpy—universally symbolized by G (free energy, from Gibbs), S (entropy, probably from Sadi Carnot, the French engineer who had first studied the working of steam engines in abstract terms), and H (enthalpy, from Hess's law).

Periodicity

The biochemist and science writer Isaac Asimov once pointed out a curious parallel in the histories of 19th-century inorganic and organic chemistry. In both cases chemists seemed to be overwhelmed by their materials. In inorganic chemistry there seemed to be too many elements, to the evident concern of Prout and others; in the case of organic chemistry there were too many compounds, as Wöhler implied in his famous remark about the subject being a jungle.

Order and coherence was given to organic chemistry by the ideas of valence and molecular structure in the decades between 1850 and 1880. In these same decades order was given to inorganic chemistry by the Germany-trained Russian chemist Dmitri Mendeleev (1834–1907). Both of these orderings hinged on the recommendation of Cannizzaro at the first ever international congress of chemists at Karlsruhe in 1860: that all chemists should accept Avogadro's hypothesis that equal volumes of gases at the same temperature and pressure contained an equal number of molecules. Accepting such a rule provided an agreed list of atomic weights that provided unique empirical formulae for organic chemists, which could then be used in determining structural formulae, and weights that were to produce Mendeleev's periodic table.

Although many precursors had tried to find order among the sixty or so known elements, it was while Mendeleev was casting around for a way of structuring a chemistry textbook in 1869 that he hit upon the periodic law. According to this, when elements are arranged according to the magnitude of their atomic weights, they display a step-like alteration in their properties such that chemically analogous elements like the alkali metals (lithium, sodium, potassium, etc.) and the halogens (fluorine, chlorine, bromine, iodine) fall into natural groups.

Philosophers of chemistry have argued over whether the iconic periodic table was accepted because it was predictive of unknown elements or simply because it accommodated the known elements in an elegant and memorable way rather like the 18th-century affinity tables. It was certainly the case that chemists paid little attention to it until Mendeleev's predictions of unknown elements such as gallium and scandium were substantiated in 1875 and 1879 respectively.

Signature tunes

In 1666, Newton made the famous demonstration with a prism that white light consisted of light of different colours. It had also been long known, chiefly by makers of fireworks, that many salts burned with a coloured flame. However, it was not until the 19th century that serious attempts were made to use flame colours to identify the presence of particular metals or elements in a sample of an unknown substance.

It was while exploring the use of coloured flames to identify the presence of elements in springs around Heidelberg that Robert

Bunsen designed the gas burner that is named after him. It was designed to produce both a non-luminous flame for flame analysis, as well as a normal flame for combustion. As a senior professor at Heidelberg, Bunsen was successful in persuading the Baden government to appoint Gustav Kirchhoff (1824–87) to a chair of physics. It was Kirchhoff who suggested to Bunsen that he could refine flame analysis by looking at the flames through a prism, which would break the light into a characteristic spectrum.

The new instrument, which they named a spectroscope, proved to be a powerful new analytical instrument—one that in various shapes and forms continues to be probably the most important instrument in contemporary chemistry. When Bunsen and Kirchhoff announced the perfection of the instrument in 1859, they were able simultaneously to reveal that they had detected and isolated two new elements with it: caesium and rubidium. Within five years two other elements, thallium and indium, had been discovered through spectroscopy, and the instrument also began to revolutionize astronomy and create a new subdiscipline of cosmochemistry.

It was now possible to detect elements in the sun, the stars, and distant nebulae, and even to detect the presence of an unknown element in the sun that the British astronomer Norman Lockyer (1836–1920) named helium in 1868. This was first example of analysis at a distance where the chemist no longer directly handled the substance that was being identified.

The fact that individual elements had such distinctive and often complex spectra—their signature tunes—also raised theoretical questions concerning their meaning and cause. Once again, Prout's hypothesis raised its head. Could one detect patterns of elements of lower atomic weight in the spectra of heavier elements, as Lockyer suggested in the 1870s after studying solar spectra? Were heavier elements disintegrating or 'dissociating' into elements of lower atomic weight in the hot atmospheres of the stars? Why, in the language of the time, did some lines (the principals) stand out? Why were some lines sharper, or more diffuse, than others? (This terminology was to continue into 20th-century quantum chemistry in the notation of s, p, d, and f orbitals.) Were there mathematical patterns to be found in the distribution of spectra among the elements?

Many bizarre Pythagorean papers on such topics were published, though it was not until a Swiss school teacher, Johann Balmer (1825–98), pointed out in 1885 that the first four lines of a hydrogen spectrum could be expressed in a simple equation that linked wavelengths in centimetres with a constant and a simple arithmetical series. This observation, which was clarified by the Swedish physicist Johannes Rydberg (1854–1919) in 1890, remained a curiosity until other spectroscopists, who were exploring the hydrogen spectrum in the infrared and ultraviolet, discovered new patterns that corresponded to simple variations of the Rydberg formula series.

None of this made much sense until the Dane, Niels Bohr (1885–1962), building on the simple model of the planetary atom of electrons rotating around a nucleus, showed that, when reckoned by the new quantum theory of Max Planck (1858–1947), the energy lost when an electron transferred from one outer orbit to an inner one would follow a Rydberg series.

In the hands of Bohr, spectra turned out to be expressions of the electronic structures of atoms and molecules. By the mid-1920s, spectroscopy had integrated a good deal of physical and inorganic chemistry. Using the periodic table as his guide, and earlier speculative models of atomic structure, Bohr was able to assign orbital structures to each of the known elements in such a way that provided for the first time an explanation of their characteristic valences. Thus the alkali metals in Group 1 of the periodic table all possessed a single electron in their outermost orbits (e.g. lithium, 2,1; sodium (2,8,1, etc.), whereas the halogens in Group 7 lacked a single electron that would complete an orbit (e.g. F, 2,7; Cl, 2,8,7, etc.). The rare gases were inert because their outermost electronic orbits were completely filled.

Ideal gases

Robert Boyle is still best remembered by the scientific community for his demonstration that the volume of a sample of atmospheric air is inversely proportion to the pressure exerted on it. In other words, the volume decreases as the pressure is increased.

Boyle's law, as it became known, was found to be true for the individual gases as they were isolated during the 18th century. By the 19th century, following the development of mathematical analysis, it was possible to state Boyle's law in the form of an equation, $pv = k$, where p and v are the respective pressures and volumes, and k is a constant. Between 1780 and 1802 French experimentalists also showed how the pressures and volumes of gases were dependent on temperature (Charles and Gay-Lussac's law)—indeed, it was intuitively obvious that gases expanded on heating. In 1832, during his investigations of the physics lying behind the behaviour of steam engines, the French engineer, Émile Clapeyron (1799–1864), combined the two gas laws into what became known as the ideal gas law, $pv = kt$, where t is the temperature.

Although Newton had provided a rational explanation of Boyle's law in terms of repulsion between the particles that made up a volume of air or a gas, it was not until the 1850s that a sophisticated kinetic theory of gases emerged in the hands of Clausius. By imagining the particles of gas to be in motion, he suggested that temperature was purely a function of the kinetic energy of the collective particles, and by using data published by the French experimental chemist Henri Regnault (1810–78), he formulated what he described as the general equation for an ideal gas, $pv = Rt$, where t is the temperature expressed in the absolute temperature scale that had been suggested as a standard by the Glaswegian physicist William Thomson (1824–1907) some years before. (He gave the symbol R for the gas constant both in tribute to Regnault and to Clapeyron, who had already used the symbol to stand for *ratio*.)

The gas analogy

The effects of passing an electric current through a solution (electrolysis) had intrigued chemists ever since Alessandro Volta (1745–1827) described the properties of his battery in 1800. Electrolysis had led to the development of Berzelius' overarching electrochemical theory that dominated inorganic and organic chemistry until the 1850s.

What exactly happened during electrolysis was the subject of one of Michael Faraday's most important investigations during the 1830s. Faraday (1791–1867) was the first chemist to subject electrolysis (a word coined on Faraday's behalf by the Cambridge polymath, William Whewell, in 1831) to quantitative investigation by establishing that the amount of a chemical compound that was decomposed in solution was directly proportional to the amount of electricity that was consumed. Thus, when copper sulphate solution was subjected to a current of electricity, the amount of copper deposited at the cathode (Faraday's term for the negative end of a circuit) depended on the amount of electricity consumed.

By comparing the weights of electrochemical deposits produced from the same amount of electricity, Faraday was also able to show that they were in the proportion of their equivalent or atomic weights. This generalization not only gave a new way of determining equivalent weights, but confirmed that the nature of matter was intimately connected to electricity. Following Berzelius' theory of salts, Faraday also supposed that the current dissociated the molecules of copper sulphate into positively charged copper and a negatively charged 'sulphate' group. Adopting Whewell's suggestions, he named these charged molecules *ions*, and *anions* and *cations* according to whether they were positively or negatively charged. There was no assumption that these ions were already present in the solution.

During the 1850s, Clausius, as a result of his development of a kinetic theory of gases, suggested that in both the liquid and gaseous states, molecules continually bombarded one another and exchanged positions. In asserting this he drew upon Williamson's work on etherification, in which radicals changed position in double decompositions. Clausius was the first chemical physicist to suggest that there was a partial dissociation and ionization of electrolytes.

Following his brilliant demonstration that stereoisomers in organic chemistry depended upon the fact that organic groups were arranged around a tetrahedron, van 't Hoff turned his attention to physical chemistry and the age-old problem of affinity. Was there some way to measure forces of affinity? Perhaps affinity was related to osmotic pressure—the force that had to be applied to prevent a solvent from diffusing through a membrane. Botanists, who were naturally interested in osmosis as a phenomenon of plant life, had previously developed a convenient apparatus for determining osmotic pressures, and van 't Hoff set about using membranes to measure the affinity that salts had for water—the water of crystallization of solid salts.

It was while making his measurements that it suddenly struck the Dutch chemist that there was an analogy between osmotic pressure and the pressure component of the ideal gas equation. Whereas in the gaseous state pressure was inversely proportional to the volume, in the case of solutions the osmotic pressure was directly proportional to the concentration of solute (that is, inversely proportional to the volume). It followed that osmotic pressure obeyed the gas law $pv = nRt$ (where n is a measure of the number of molecules present), and that at equal osmotic pressure and temperature, equal volumes of solutions contain an equal number of molecules—exactly the same number which, at the same temperature and pressure, is contained in an equal volume of gas. If this sounds suspiciously like Avogadro's hypothesis, it is, and the analogy between the behaviour of a perfect gas and the solute in a dilute solution was complete.

Further experimental results showed that there were considerable deviations from the gas law unless the dilution was great enough for the volume occupied by molecules to be ignored. Even then, inorganic salt solutions gave much higher readings of osmotic pressure than the gas law predicted. A similar disparity had been noticed in measurements of freezing point depressions when identical amounts of different salts were added to water. Undismayed, van 't Hoff's solution was to add an empirical correction factor, i, into the gas equation, $pv = iRt$. Curiously, it seemed that the i factor always had to be greater than one; for example, in hydrochloric acid it was 1.98, and for sodium sulphate 1.82. These findings and suggestions were published in 1885 and noticed by a young Swedish doctoral student.

Ionic theory

Svante Arrhenius (1859–1927) was the son of an estate manager, born at Wijk near Uppsala, Sweden. He read chemistry and physics at the local university before studying the properties of electrolytes at the University of Stockholm for his doctorate in 1884. The thesis was not well received, making it difficult for him to obtain an academic appointment for some years.

In 1887 he developed the dissociation theory. According to this, in a weak electrolyte, molecules were already dissociated (ionized) before a potential difference was applied to them. The theory, according to which dilute solutions obey a general gas law, $pv = iRt$, where i is a measure of dissociation, tied in beautifully with van 't Hoff's views on osmotic pressure. Together Arrhenius, van 't Hoff,

and the German Wilhelm Ostwald (1853–1932) launched the
Zeitschrift für physikalische Chemie in 1887.

Despite difficulties with strong electrolytes (whose behaviour was
not satisfactorily explained until the 1930s), the ionic theory
found rapid use among chemists and biochemists. In 1889
Arrhenius applied van 't Hoff's equation for the temperature
dependence of chemical equilibria to the kinetics of chemical
reactions, showing that reactivity depended upon what he called
'activation energy'. The notion that molecules require a certain
critical energy to react has proved fundamental in understanding
reaction mechanisms within living systems.

Arrhenius was awarded the Nobel Prize in Chemistry in 1903
for his theory of electrolytic dissociation, which also enabled
biochemists to develop the pH scale of hydrogen concentration
and to understand how living tissues 'buffered' themselves
against excessive acidity or alkalinity—the basis of homeostasis
in living systems. Arrhenius became an international figure in
science. His originality lay in an ability to move between fields
on the boundaries of established disciplines—something that
many later chemists proved good at.

Kinetics

In his dispute with Proust at the end of the 18th century,
Berthollet had argued against compounds being formed in
definite proportions on the grounds that if one reactant was
present in greater amounts it would inevitably combine with the
other reactant in variable amounts. What Berthollet was thinking
was that the respective masses of reactants were bound to affect
the course of combination. In the 1850s this law of mass action
was taken up by Williamson, who suggested reactions were
frequently in dynamic equilibrium—in which case, if one reactant
was increased in concentration the overall reaction would shift in
a particular direction.

This concept was explored by a Norwegian chemist, Cato Guldberg (1836–1902), and his mathematical colleague, Peter Waage (1833–1900), in 1863; they expressed Williamson's idea in terms of the concentrations of the reactants and the speed of formation of products. If the forward reaction was twice as fast as the reverse reaction, the equilibrium was far to the right; conversely, if the reverse reaction was twice the rate of the forward reaction, the equilibrium would shift to the left. In both cases the rates depend solely on the frequency of collision between molecules. By making the concentration of the reactants greater, the forward rate would increase.

Guldberg and Waage showed, using square brackets to represent concentrations, that [C][D] divided by [A][B], which represent the concentrations of the reverse and forward reactions respectively, will be a constant they named as the equilibrium constant. This constant came to play a major role in understanding chemical kinetics and the mechanisms of chemical reactions. However, because they wrote in Norwegian it was not until the 1880s that their work became noticed and utilized.

The changing laboratory

The principal problem presented by chemical work is fairly obvious. Chemists produce obnoxious and sometimes dangerous fumes (British school children have nicknamed it 'stinks' ever since the 1870s). Laboratory atmospheres need ventilation beyond that supplied by open windows. The latter might suffice for one experimenter, but not for a gathering of twenty to thirty students.

Prior to the 1860s, the usual way of dealing with these ventilation problems was either (as Liebig did) to use window ventilation backed up by chimney hoods, or to put a laboratory in the basement adjacent to the main drains and chimney (as at the Royal Institution and, later, at the Pharmaceutical Society). The Pharmaceutical Society's laboratory (1842) was actually built in the back garden of a house in Bloomsbury Square, with skylights

Reactivity

105

and sliding vents leading to the main chimney where evaporations could be conducted. Drains from the sinks led directly into the town sewers. Hydrogen sulphide gas was also on tap and in its own ventilation shaft.

The Society's design worked well enough for the chemists in it, but drifting smells made life uncomfortable for those in the rooms above. Consequently, in the 1860s the laboratory was moved to the top storey of the building, and where there were laboratories for several scientific disciplines this became the preferred solution: put the chemists next to the sky! Even so, by the 1870s it was clear that more powerful ventilation systems were necessary, and using electric fans on the roof that sucked air through the ventilation shafts solved the problem. Ventilation hoods, sand baths, drying ovens, and water supplies for filter pumps became essential design specifications (see Figure 14).

14. Birkbeck Laboratory at University College London. The foreground area was mainly used by University College School for demonstrations. From a lithograph of 1875.

The rise of physical chemistry at the end of the 19th century brought further changes to laboratory design. Instruments and techniques such as ultraviolet and infrared spectroscopy, spectrophotometers, X-ray cameras, mass spectrometers, and chromatography began to offer new, simpler, and faster methods of analysing materials and directly assigning a structure to them. Such physical methods had completely replaced wet and dry methods of analysis by the 1960s, so that today the traditional laboratory bench of Liebig's and Kolbe's time has lost its scaffold of reagent shelves and returned to a table-like appearance.

The inert gases

The unexpected discovery of the so-called rare or inert gases by the physicist Lord Rayleigh (1842–1919) and the chemist William Ramsay (1852–1916) in the 1890s, and Ramsay's placement of them in the periodic table as a new group, also did much to substantiate the usefulness of the periodic law. Serious puzzles remained of course, notably where to place the large number of so-called rare earth elements (lanthanides) which had virtually identical properties. A solution had to await new insights into the structure of the atom that the discovery of X-rays, radioactivity, and the electron produced from the 1890s onwards.

Despite attempts by Ramsay and many other chemists to form compounds with the rare gases, they remained obstinately inert. Hitherto all elements and compounds known to chemists displayed transformations and transmutations, but these gases were totally unreactive. Chemists were, of course, familiar with the relative inertness of nitrogen as a gas, or gold and platinum among metals, or the paraffins among organic compounds. But given a hammer blow from heat, electricity, or pressure, such substances were always transformable.

Argon refused to react no matter what physical conditions to which it was subjected. To make matters worse, when its specific gravity was determined it was clear that argon was monatomic, unlike any of the common elementary gases. The explanation for the inertness of argon and its congeners was forthcoming only when Bohr developed the model of electronic shells in which the inert gases had full complements of electrons in their outer shells. As Linus Pauling (1901–94) pointed out in 1932, this completeness need not necessarily prevent compound production if an inert gas was reacted with a sufficiently energetic reactant such as fluorine. Curiously, Pauling's observation went unheeded and the first of the inert gas compounds, xenon fluoride, was not prepared until 1962 by Niels Bartlett (1932–2008).

Social chemists

Despite the lead shown by first France and then Germany in the development of chemistry in the early 19th century, it was the British who first organized chemists into a Chemical Society. The London Chemical Society (now the Royal Society of Chemistry) was formed in 1841, largely in response to the development of chemistry on the continent and the feeling that British chemists were being left behind, particularly in organic chemistry. The French followed in 1857 with the Société chimique de France. On his appointment to a chair at the University of Berlin in 1865 Hofmann, who had been the leading light of the London Chemical Society, organized the Gesellschaft Deutscher Chemiker. Other countries followed swiftly, the most important being the formation of the American Chemical Society in 1876 as a result of an earlier commemoration of Joseph Priestley at his former home in Pennsylvania.

Each of these organizations began to publish proceedings and transactions in their own journals which, in turn, helped to forge national chemical communities, so that by the 1880s there was a proliferation of literature for the chemist to follow in a variety of languages and in a variety of specialisms. Abstracts and translations became necessary, as well as guides and surveys to what had been already published, such as the serialized handbooks of Gmelin and Beilstein (see Chapter 4).

International meetings of chemists, made possible by the transformation of transport that decreased the tyranny of distance, were initially held to decide upon standards. These began in 1858 when Kekulé and others called chemists to Karlsruhe to adjudicate on standards for atomic weights, and this was followed by the meeting in Geneva in 1892 to formulate rules for organic nomenclature (Chapter 4).

The relentless increase in specialization soon led to conferences arranged around particular subjects within the categories of analytical, inorganic, organic, and physical chemistry. For example, in 1923 the Faraday Society, a society specializing in physical chemistry founded in London in 1903, held a meeting on the electronic theory in chemistry. Overseas study, travel to conferences, and socializing in different academic and industrial contexts, meant that 20th-century and contemporary chemists form what the Swedish historian Staffan Bergwik has called 'capillary networks through which the physical sciences migrate through laboratories, universities, academies and industries'.

In answering the questions 'What changes in a reaction?', 'How fast do changes take place?', and 'How complete are the changes?', physical chemists made chemistry much more mathematical. This was the first of many substantial alterations that the discipline of chemistry underwent as the subject moved into the 20th century.

Chapter 6
Synthesis

The shape and scale of chemistry was shaken by a series of transformations in the 20th century: two world wars; a shift from coal to oil as the feedstock for chemical industry; the introduction of physical instrumentation, quantum mechanics, and electronic theories; and the organization of academia and industry to create Big Science as opposed to the more individualized research of previous centuries.

Underlying these important changes was the theme of synthesis of natural chemicals and the creation of artificial materials. Accompanying the transformations that the discipline of chemistry underwent, there was a shift from European dominance of the subject to the US and, from the 1960s onwards, the emergence of new centres of research in the Soviet Union, Japan, and China. Closely connected with this globalization was the impact of two world wars and the so-called 'Cold War', which effectively made research sponsored by governments and the military the paymasters of most chemical work just as it did in the world of physics.

Another dramatic change was that what had been a male-dominated profession saw women engage with chemistry in ever increasing numbers. Within chemical practice itself, there was an instrumental revolution after 1950 which saw the virtual disappearance of dry and wet analyses and their replacement by electronic machinery; by the same token, the way that chemistry was taught and the way textbooks were written were necessarily transformed.

Finally, although historically chemistry had always had a close affiliation with pharmacy and medicine, the subject made a decided shift towards the biological sciences after World War II, as well as becoming what some commentators have seen as a service science to other areas of the physical sciences as the neologisms astrochemistry, geochemistry, and materials science may imply. To others, however, this was seen as the infiltration of chemistry into the other sciences and chemistry becoming the fundamental basis—the central science—for the study of nature.

The beginnings of organic synthesis

Alchemists, chymists, and chemists have always been involved in imitating and improving nature. According to the OED, the first recorded use of the term *synthesis* occurred in the lectures of Peter Shaw, an English Newtonian lecturer and disciple of Boerhaave. Lavoisier successfully analysed and synthesized both 'air' and water. He commented in his *Elements of Chemistry* (1790) that chemists must never be satisfied that they know the composition of a substance unless it has been both analysed and synthesized—something that was implicitly built into the 18th-century tables of affinity.

By the beginning of the 19th century the classification of materials as animal, vegetable, and mineral had become simplified into the duality of mineral/inorganic and organic/organized. Wöhler's artificial preparation of urea (see Chapter 4) confirmed the phenomenon of isomerism and that organic compounds obeyed the same Daltonian laws and that their construction was possible.

But how could the chemist determine the ways in which atomic arrangements differed in isomers and in organic compounds generally? The answer was through *metamorphosis*, or the analysis of degradation of organic compounds into simpler products whose compositions were known with some certainty. This is why many early papers on organic compounds are entitled 'on the metamorphosis of x, y, z, etc.' In effect, chemists used the term metamorphosis instead of synthesis and hoped that the isolation and identification of common (or very similar or related) degradation products would lead to an understanding of parental composition.

Liebig and his school at Giessen became masters of this form of 'analytical synthesis' or 'synthetical experiments' by exploiting the accurate quantitative analyses made possible by the *Kaliapparat*, improved purification techniques, and the Pythagorean possibilities of paper chemistry based on stoichiometric deductions. The task facing chemists was challenging. The empirical formula of quinine $C_{20}H_{24}N_2O_2$ tells one nothing about how the atoms are arranged; it is like being told to find an intelligible word (quinine) made up of the letters N_2QEUI_2.

By the 1830s, however, chemists had Berzelius' electrochemical theory and Gay-Lussac's concept of radicals to guide them. By using empirical molecular formulae, the analysis of decomposition products, and radical and electrochemical theories, Liebig and his successors—notably Hofmann working in London at the Royal College of Chemistry—began to deduce compositional arrangements. For example, the metamorphoses of benzaldehyde and uric acid suggested that the radicals *benzoyle* and *uril* provided a common constitutional arrangement. Liebig and Wöhler's one hundred-page, 1838 paper on uric acid identified some sixteen new derivatives.

Such successes led Liebig and Wöhler to announce boldly in 1838 that:

> The production of all organic substances no longer belongs just to organisms. It must be viewed as not only probable, but as certain, that we shall produce them in our laboratories. Sugar, salicin, and

morphine will be artificially produced. Of course, we do not yet know how this goal can be achieved because we do not yet know how the precursors out of which these compounds arise. But we shall come to know them.

In London, Hofmann laid down a vocabulary of reagents and reactions that enabled chemists to be more confident about arrangements of atoms within molecules. In a paper on toluidine, published in 1843, Hofmann declared that the time was drawing near when the information gleaned from metamorphoses and analytical experiments would be sufficient to enable chemists to construct valuable commodities like quinine. Naphthalene (he pointed out in 1849) is abundantly available from coal gas production and is easily transformed into an alkaline base, naphthalidine, $C_{20}H_9N$ [C=6, O=8]. But if quinine was $C_{20}H_{11}NO_2$, the sole difference was two equivalents of water [HO]. Hofmann was not so naive as to hope that adding water would achieve a synthesis; instead he forecast that eventually an appropriate metamorphic reaction would be discovered.

With aspirations such as these pervading the laboratory at the Royal College of Chemistry, it is scarcely surprising that 18-year-old William Perkin (1838–1907) should attempt to transform allyltoluidine into quinine in 1856 by removing hydrogen and adding oxygen. When this experiment failed he tried aniline instead, and serendipitously created the British dyestuffs industry with the beautiful mauve dye that resulted. Hofmann was to follow suit on his return to Germany in 1865 and create the complementary and eventually world-leading German industry.

Frankland recognized that the organometallic compounds he had prepared would be powerful aids to synthesis, by which he meant the chemist's ability to build up new compounds 'stone by stone' with a view to understanding their atomic configurations. His German friend Kolbe had already shown in 1845 that there really was no difference between inorganic and simple organic compounds, by accidentally constructing acetic acid from its elements via carbon disulphide and tetrachloroethylene intermediates. Between 1863 and 1870 Frankland exploited zinc ethyl and other organic reagents, including ethyl acetate, in the construction of ethers, and dicarboxylic and hydroxyl acids. This meticulous work revealed clearly the structure and relationship of these compounds, while its methodology had great bearing on the growth of the chemical industry.

In determining chemical structures, chemists soon became adept at manipulating molecules by using reactions that followed regular patterns. For example, to increase a carbon chain by two atoms of carbon, two different aldehydes (CHO group) are heated together in the presence of a sodium salt of a fatty acid; the two aldehydes join together with the elimination of water. This became known as Perkin's reaction or condensation, after it was devised by Perkin to make cinnamic acid. Hofmann devised a reaction whereby a ring compound containing nitrogen could be sprung apart to form an open chain, and this became known as a Hofmann degradation. By the 20th century, organic chemists were taught and memorized hundreds of these 'named reactions', all of which could be exploited in planning syntheses.

By 1860, Marcellin Berthelot felt able to publish a book entitled *Chimie organique fondée sur la synthèse* and to lay down rules for the progressive synthesis of organic compounds from first principles by 'the sole aid of chemical means' (i.e. no vital forces were invoked). However, his yields (like those of Kolbe before him) were ridiculously low and it took another decade of structural modelling, laboratory building, and the development of new glass apparatus and experimental techniques before the hopes and dreams of Liebig, Hofmann, and Berthelot concerning useful (and industrially exploitable) synthetic pathways began to be fulfilled.

The first half of the 20th century was dominated by the determination of the structures and synthesis of naturally occurring compounds for which the Nobel Prizes instituted in 1901 were frequently the reward. Two important exemplars were the German chemists Baeyer and Fischer. The son of a Prussian military officer, Adolf von Baeyer (1835–1917) was born in Berlin and studied chemistry with Bunsen and Kekulé at the University of Heidelberg. He taught organic chemistry in Berlin and Strasbourg before succeeding Liebig at the University of Munich in 1873, where he spent the rest of his career (see Figure 15).

In 1905 he was awarded the Nobel Prize in Chemistry for his 1880 synthesis of the natural dyestuff, indigo. The synthesis, which was subsequently scaled up commercially by the Badische Anilin- und Soda-Fabrik (BASF), destroyed the economy of agriculturally produced indigo and confirmed the ability of organic chemists to synthesize natural products.

Baeyer's pupil, Emil Fischer (1852–1919), held chairs at the universities of Erlangen and Würzburg, before spending the

15. Adolf Baeyer with his University of Munich students in 1877.

remainder of his life from 1892 at the University of Berlin and the newly established government-funded Kaiser Wilhelm Institute. Fischer extended the work of Baeyer by the synthesis of complex organic compounds that enabled him to establish the exact constitutions and structures of sugars, enzymes, purines, and proteins. He thereby laid the chemical foundations for biochemistry and the study of macromolecules, and established a methodology of degradation and synthesis in natural products chemistry that remained in use until the 1950s, when physical and instrumental methods of structure determination became firmly established.

His 'lock and key' model of enzyme action was announced in 1894 and proved extremely fruitful. In 1897 Fischer established that biologically important molecules such as uric acid, xanthine, adenine, and guanine were related to a heterocyclic nitrogenous base he named purine. Subsequently, he synthesized many purines, including the barbiturate that was exploited commercially as Veronal. From 1899 onwards Fischer turned to the structure of proteins, showing them to be composed from amino acids grouped in polypeptide chains. By 1916 he had synthesized over a hundred polypeptides.

Fischer was the first German to be awarded the Nobel Prize in Chemistry in 1902—in recognition of his syntheses of sugars and purines. With his black beard and pince-nez, Fischer was an awesome figure in the laboratory which he ruled like an army general. His over 350 research students included the chemistry Nobel Prize winners, Fritz Pregl (1923), Adolf Windaus (1928), Hans Fischer (1930), and Otto Diels (1950); and for medicine, Karl Landsteiner (1930) and Otto Warburg (1931). Despite ill health caused by the dangerous chemicals he worked with, he was an active patriot during World War I, organizing Germany's chemical and food supplies. His disillusionment following the death of two of his sons on the battlefront caused his suicide in 1919.

The American Robert Woodward (1917–79), working at Harvard, was the most prominent synthesizer in the mid-20th century (see Figure 16). He triumphantly synthesized quinine (1944), cholesterol and cortisone (both in 1951), chlorophyll (1960), and vitamin B_{12} (1976). Ironically he was to receive the Nobel Prize in 1965 not for synthesis, but for his deduction of rules derived from quantum mechanics concerning the conservation of orbital symmetry. This demonstrates the way physical chemistry had transformed organic chemistry since the 1930s and deeply affected the way viable synthetic pathways were sought by organic chemists.

Woodward's European rival was the Glaswegian, Alexander Todd (1907–97), who was awarded the Nobel Prize in Chemistry in 1957 for his elucidation of the structures and syntheses of nucleosides, nucleotides, and their co-enzymes, as well as the related problems of their phosphorylation. This work was a necessary preliminary to the elucidation of the structure of DNA by the non-chemists James Watson (a zoologist) and Francis Crick (a physicist) in 1952. Todd also worked extensively on adenosine triphosphate, vitamin B_{12}, and other natural products, including plant and insect pigments. A towering physical presence in organic chemistry, he was nicknamed 'Lord Todd Almighty'.

These triumphant structural determinations of natural products were surpassed by the ability of 20th-century chemists to create new, unnatural substances, beginning with plastics and resins, and polymers such as rayon and nylon. On the other hand, the ability to synthesize a molecule was not necessarily useful. For example, amid wartime pressure to treat dangerously infected wounds, British chemists made great efforts to synthesize penicillin, the first of the antibiotics, but it proved too difficult to scale up production; instead penicillin was manufactured by culturing the mould directly, as were other antibiotics whose structures were known such as streptomycin and tetracycline. This refashioned brewery technology soon became known as biotechnology.

16. Robert Burns Woodward (1917–79), America's most successful organic chemist.

Chemistry, as the science of transmutations and substitutions, turned the natural world upside down. The imitation of nature led to the possibility of breaking with Nature and surpassing it to form an artificial world. However, none of this would have been possible without the physicists' analysis of the deep structure of the atom that had begun at the end of the 19th century.

The electronic age

Research in the 19th-century on the conduction of electricity was marked by a series of brilliant and intensely visual investigations by German and British experimentalists. The two most outstanding investigators were Johann Hittorf (1824–1914) and William Crookes (1832–1919), whose research was made possible by the innovations and improvements in glass blowing and vacuum pump technology instituted by the Bonn instrument-maker, Johann Geissler (1815–79).

Hittorf, a physical chemist, investigated how gas spectra altered under changing pressure. It was during his experiments to map gaseous discharge and to quantify the variation of electric force and temperatures that he discovered shadows on the phosphorescent patches glowing on the discharge tubes. From this he inferred that ions were emitted from the cathode, though he left it to others to develop the notion of cathode rays.

By then (1879) Hittorf's work had been completely overtaken by Crookes, who had approached gaseous discharges from a completely different direction, namely the puzzle of what went on inside a discharge tube. Geissler's technique for evaluating the quality of a vacuum by residual glow during conductivity encouraged Crookes to interpret his work as evidence for a fourth state of matter he called 'radiant matter'. The Crookes tube, as his form of discharge tube became known, not only allowed Conrad Röntgen (1845–1923) to uncover X-rays in 1895, but paved the way for J. J. Thomson's postulation of the electron in 1897, thus opening the way for models of the atom in terms of a nucleus surrounded by electrons.

Once Thomson had shown that the electron was a subatomic particle, he spent a decade investigating heavier positive particles that were also produced during the discharge of electricity through

Synthesis

gases. Together with his assistant, Francis Aston (1877–1945), in 1912 they constructed a mass spectrograph, using a camera to record spectra. The positive ray spectra produced in this machine proved to be a superb way of identifying and distinguishing ions and the different isotopes of elements. With many improvements over the next fifty years, the mass spectrometer (as it was renamed) became a major analytical tool not only for chemists but in almost every area of scientific and industrial activity.

One of the anomalies Mendeleev had to deal with in the periodic table were 'pair reversals', where a group of elements with similar properties contained an element with an atomic weight that was higher than the element that followed in the atomic weight sequence. For example, the atomic weights of tellurium and iodine were 128 and 127, but if iodine was placed before tellurium in the periodic table it would be displaced from the halogen Group 7 where it clearly belonged. To Mendeleev this suggested that the atomic weight of tellurium was probably incorrect.

The reason for pair reversal only became apparent when knowledge of isotopes was suggested by Frederick Soddy (1877–1956) in 1912. Some elements such as tellurium possessed up to half a dozen isotopes, including an abundant one of mass 128. This mass had the effect of creating an average atomic weight greater than that of iodine which has only one isotope of mass 127.

Around the same time the Oxford physicist Henry Moseley used the frequency of the most intense line in an element's X-ray spectrum to show that there was a more fundamental way of ordering elements, namely their atomic number or the size of the nuclear charge. Moseley's work, coupled with that of Soddy and Aston on isotopes, revealed the reason for reversal anomalies in Mendeleev's periodic table. Before his untimely death at Gallipoli in 1915, Moseley also demonstrated the probable existence of a number of undiscovered elements, which were gradually identified over the following decades.

With his first instrument Aston was able to announce that neon was a mixture of two isotopes; and within a few months he had analysed a sufficient number of other elements to be able to announce a 'whole number rule' according to which, if atomic masses were calculated with respect to oxygen (mass 16) then they were all integers, just as Prout had suggested. He concluded that atomic weights determined by chemical methods were merely 'fortuitous statistical effects' caused by the relative quantities of the isotopic constituents. In the case of neon, the relative abundances of Ne^{22} and Ne^{20} (a nomenclature introduced by Aston) produced a non-integral atomic weight of 20.1. Was Prout's hypothesis true after all?

The problem here was that hydrogen was anomalous. With a mass number of 1.008 it did not obey the whole number rule. The solution was found in Einstein's special theory of relativity and the famous equation linking energy, mass, and the velocity of light: $e = mc^2$. The American maverick chemist, William Harkins (1873–1951), suggested in 1915 that atoms might be considered to be helium ions (nuclei) surrounded by electrons. The helium 'atoms' were so closely packed together in elementary atoms that they lost mass in the form of energy.

Aston seized upon this packing concept in 1920 to explain the hydrogen anomaly, but used Prout's original idea of hydrogen atoms forming other elements rather than helium. Suppose elements were constructed from hydrogen atoms of individual mass 1.008. Mass was only additive, he argued, when nuclear charges were relatively distant from one another. In the case of atoms other than hydrogen, mass was lost within the atom in the form of a binding energy as the individual atoms of hydrogen were squeezed or packed together.

As Aston slowly improved the accuracy and sensitivity of the mass spectrometer through the 1920s, it became apparent that the whole number rule was not exact and that there were many deviations. Aston codified these deviations by defining a new function called the 'packing fraction', defined as the divergence of an atomic mass from a whole number divided by its mass number. By plotting these fractions against mass numbers Aston obtained a simple curve, which threw important light on nuclear abundances and stabilities. The packing fraction, along with atomic mass and atomic number, soon became the three defining atomic constants for both chemists and physicists, and Aston was awarded a Nobel Prize in 1922 for his work on isotopes and the whole number rule.

Making organic chemistry physical

Although 19th-century organic chemistry became a coherent subdiscipline based upon knowledge of bonding and structure, and inorganic chemistry had come to order through the periodic law, chemistry remained essentially an empirical science of analysis and synthesis based upon centuries-old techniques of extraction, distillation, etc. It was for this reason that Ernest Rutherford could dismiss chemists as mere stamp collectors compared to physicists like himself, who were 'true' scientists. Ironically, this was to change largely because of Rutherford's own work in revealing the structure of the atom. Atomic structure explained why the periodic table worked, as well as introducing new ideas concerning why organic reactions proceeded in the way they did.

Three important innovators in physical organic chemistry were the two English chemists, Arthur Lapworth (1872–1941) and Christopher Ingold (1893–1970), and the American Louis Hammett (1894–1987). Lapworth was the transitional figure who linked 19th-century structural theory with models of electronic valence. Interestingly, Lapworth married one of three musically gifted sisters who each married a chemist—the other two married the organic chemist, W. H. Perkin Jr (1860–1929), and the pioneer of silicon chemistry, Frederic Kipping (1863–1941). The marriages are a reminder that chemists are often found to be talented musically, the most famous example being the Russian organic chemist, Aleksandr Borodin (1833–87).

The 19th-century assumption that there was one unique structure for each compound was soon found wanting in the identification of tautomerism, in which a compound flips from one structure to another. As chemists became more familiar with the electron, notably after G. N. Lewis suggested that substances coupled together by completing an octet of electrons in their outer electronic shells, Lapworth developed ideas about electrons moving through double-bonded organic compounds under the influence of other polarized elements or compounds. In some ways this represented a return to Berzelius' earlier electrochemical theory.

Lapworth's ideas were adopted and extended by the Oxford chemist Robert Robinson in the 1920s, and it was he who introduced the ubiquitous and iconic curved arrow symbolism for electron movement in 1922. However, Robinson was not greatly interested in electronic theory except as an adjunct to understanding the structures of natural products, and he left it to his London rival Ingold to use the electronic theory to classify organic reaction mechanisms generally using kinetics, spectroscopy, and other tools.

Ingold introduced a kinetic terminology of substitution and elimination reaction mechanisms in a review paper in 1934, and in his subsequent magnum opus *Structure and Mechanism in Organic Chemistry* (1953). His collaboration with a Welsh colleague Edward Hughes (1906–63) became on a par with that of Liebig and Wöhler a century before, though it should be noted that much of Ingold's early theoretical work had been done with his wife, Hilda, a pupil of Martha Whiteley.

Meanwhile, at Columbia University in New York, Louis Hammett used his knowledge of physical chemistry to develop equations for acidity and rate constants that could be used in elucidating mechanistic pathways. His genius was to develop a simple way of measuring the rate constants of reactions in organic chemistry. This made it easier for chemists to determine particular mechanisms of reactions. Hammett's pioneering textbook, *Physical Organic Chemistry* (1940), although little known in Europe until a decade later, not only introduced the term, but ensured that American chemists would actually forge ahead of their European competitors, as post-war Nobel Prizes clearly indicate.

In 1946 American chemists held the first of an annual series of reaction mechanism conferences and began to take over leadership in the growing field of physical organic chemistry. European wars had led British postgraduates to abandon further studies in Germany; they now turned to America instead.

The chemists' wars

The Great War of 1914–18 was the first conflict in which European chemists were involved in both defensive and offensive research. In 1914 Germany was still the world's leader in chemical research and the supply of chemicals. Consequently, Britain had much catching up to do. The Chemical Society played its part through its library, which proved a valuable resource of information on pure and applied chemistry.

The Society (like other learned societies) found itself mired in controversy as to whether honorary German Fellows should be stripped of their membership. The Council's decision not to do so caused widespread criticism and controversy in the press; in 1916 the Council capitulated and the names of Baeyer, Fischer, Nernst, and Ostwald (to name the most distinguished German chemists) were removed. It was not until 1929 that four of the surviving names were re-elected, and that new Germans and Austrian chemists were admitted. Strangely, there seems to have been no reciprocal action by the German Chemical Society. Despite the government's action in making qualified scientists reserved occupations in 1915 (following the tragic death of the chemist Henry Moseley that year), some eighty-three British chemists lost their lives on the battlefields of Europe.

The war that erupted in 1914 is popularly known as 'the chemists' war' because of the use of poisonous gas warfare. In fact, less than 1 per cent of war casualties were directly attributable to the use of chemical warfare, and the phrase was first used in 1917 to describe the way chemistry had been marshalled to place Great Britain on a war footing. By analogy the 1939–45 war has been called 'the physicists' war' because of the effort that was placed upon making an atomic bomb. In fact, chemists were closely involved in the separation of uranium isotopes and in the manufacture of heavy water. And without their skills there would have been no bomb.

In both world wars the chemical industry was dominated by the drive to improve and raise production levels of conventional high explosives and metal production for cannon. The petroleum soap (napalm) devised by the Harvard chemist Louis Fieser (1899–1977) killed more Japanese than the atomic weapons that destroyed Hiroshima and Nagasaki combined. Its use during the Korean War (1950–3) and the prolonged Vietnam war (1955–75) became a symbol of the evils of warfare and was responsible for a chemophobic swing against chemistry that has had a lasting effect on popular culture.

All 20th-century wars were chemists' wars, because the blockading of shipping and the cutting of land supply routes forced both the Allies and the Axis powers to find new ways of preparing materials that were in short supply or in inventing substitute materials. The best-known example is the scaling up of the Haber process for fixing nitrogen that could be used both to prepare nitric acid for use in explosive weapons and for the production of fertilizers to aid food production. Its application to explosives was directly caused by the Royal Navy's blockade of German shipping carrying Chilean nitre that had hitherto been the source of both fertilizer and explosives.

Although Germany was able to scale up the Haber process in time, British chemists had been left in ignorance and instead industrialized an older German process for 'fixing' nitrogen by making calcium cyanamide by passing nitrogen over calcium carbide. Ammonia was then generated from the cyanamide by exposing it to steam. Ammonia was also extracted from urban gas plants. In both Germany and Britain ammonia was then converted to nitric acid by a method that Wilhelm Ostwald had patented in 1902, whereby ammonia was oxidized in the presence of a platinum catalyst. Platinum was far too expensive to use in this process, and during the 1914–18 war German chemists found a cheaper catalyst of iron and bismuth oxides, thus enabling Germany to continue the war after 1916.

On the ammunition front, chemists had to scale up production of various aromatic organic compounds for the manufacture of TNT, amatol, and lyddite for high-explosive shells as well as smokeless cordite as a propellant. In every case chemists were met with bottlenecks as one problem produced another. For example, cordite was made by nitrating cellulose to make guncotton. This was then mixed with nitroglycerine and petroleum jelly. But to make the cordite suitably potable the nitroglycerine jelly had to be dissolved in acetone before it could be extruded into a tubular form.

However, acetone, normally prepared by the distillation of wood, was in short supply. The problem of acetone production was solved by a chemist at the University of Manchester, Chaim Weizmann (1874–1952). He developed an acetone fermentation process based upon brewery methods. Subsequently, a grateful British government signed the Balfour Agreement which gave Zionists like Weizmann a homeland in Palestine. Weizmann was one of the few chemists to have become president of a nation.

German chemists faced similar problems, as blockades that restricted cotton supplies meant that guncotton had to be prepared by an alternative method using rags and wood pulp, and glycerine prepared from sugar fermentation. Max Delbrück (1850–1919), a German brewery chemist, devised fermentation processes whereby yeast could be used to produce animal feed. These were early signs of the application of chemistry to the biotechnology that gained industrial significance after the 1960s.

Chemists also contributed to the improvement of glass manufacture, because high-quality optical glass was needed for surveying instruments in warfare; dyestuffs were another area where Britain, which had lost out to Germany (and to France to a lesser extent) at the end of the 19th century, was forced to make its own dye materials. Similar shortages of raw materials were experienced in Germany and Turkey; without direct access to Far Eastern natural rubber, German chemists explored synthetic alternatives which gave their chemists a strong start in the new area of polymer research, which not only produced many new synthetic fabrics, but also proved of long-term value in gaining an understanding of biological chemical systems.

The momentum of war-driven chemical research did not cease in 1918, nor did it in 1945. All these examples, including the use of chemical warfare, were expressions of what John Agar has termed 'organized science and managed innovation'. It is doubly ironic that the German research on chemical warfare agents was promoted in peacetime in the production of insecticides; one of these, devised by Haber and named Zyklon B, was to be used to gas Jews in concentration camps during the 1940s.

Women in chemistry

Probably the most important consequence of the 20th-century wars was that strong links were established between government, the military, industry, and academia. These continued through the short peace of the 1920s and 1930s, before becoming even more firmly established during World War II and the subsequent Cold War of the 1950s and 1960s.

Warfare in the 20th century brought new opportunities for women chemists, most of whom were drafted into fundamental bench work needed in testing reactions and synthesizing products of importance. Martha Whiteley (1866–1956), who was to be among the first twenty-one woman elected to the Chemical Society in 1920, headed a team of women at Imperial College in London that manufactured the local anaesthetic beta-eucaine, a derivative of cocaine.

Many more ordinary women were drafted into the chemical industry to perform heavy manual work in the manufacture of explosives. In the aftermath of war, the huge chemical industries built up by America, Germany, and Britain were not decommissioned and dismantled. Instead they were merged to form large and powerful companies. In Germany in 1925, BASF, Bayer, Hoechst, and other companies formed IG Farben with 100,000 male and female employees; in America Du Pont expanded into new areas of manufacture; while in Britain, Imperial Chemical Industries (ICI) was formed in 1926 from the merger of most of the surviving alkali and dyestuffs manufacturing companies.

These industrial companies provided women chemistry graduates with new opportunities for employment as they combined marriage and family with scientific careers. Until then, as male chemists' entries in biographical dictionaries confirm, they had tended to abandon research careers after marriage to male chemists they met during their training in postgraduate laboratories.

The instrumental revolution

The most 'creative toolmaker' of the 20th century was the American blacksmith's son, Arnold Beckman (1900–2004). As a teenager he became interested in how microscopy and photography could be used to study the fine structure of metals; while still in high school he founded a company to make metallurgical instruments.

Following a conventional chemical training at California College of Technology, he became struck by the fact that the productivity of the local orange-tree industry was dependent on soil acidity. The logarithmic pH scale of acidity had been introduced by the Danish biochemist Soren Sørensen (1868–1939) in 1909 (pH stood for potential hydrogen ions, with zero for extreme acidity and fourteen for extreme basicity). Measurements of pH involved elaborate volumetric analysis off site until Beckman packed everything into a simple portable meter that calculated the pH electronically (see Figure 17). The device was patented in 1934 when he left academia and founded an instrument company that not only manufactured Beckman pH meters but also absorption spectrophotometers that found a use in the growing industrialized food industry.

Beckman was in a good position when chemical practice underwent an instrumental revolution in the 1950s. Infrared spectroscopy had begun in the late 19th century when two former British army officers who were working for the government's educational department detected correlations between infrared spectra and the functional groups of organic compounds. In the early 1900s the American physicist William Coblentz published data on the spectra of over a hundred organic molecules.

However, for various reasons these findings did not lead to this form of spectroscopy being used in structural determinations. This occurred in the 1930s, when petroleum chemists found the need for a rapid way to identify the different fractions of distilled petroleum. Infrared spectroscopy then became an essential tool as the world wars encouraged a switch from coal to oil as the principal feedstock for the chemical industry. The adoption of

17. **Beckman Model G pH meter (1934). This cubic box (roughly a foot in width, depth, and height) was a voltmeter that used an acid-base responsive electrode to give a direct conversion of voltage differences to differences of pH at the temperature of measurement. It was one of the first portable laboratory instruments available. With its hidden mechanism, it was also the first chemical 'black box'.**

ultraviolet and infrared spectroscopy, nuclear magnetic resonance, and mass spectrometry transformed compounds into an abstract artwork of spectral lines or curves by an electronic machine.

There were many consequences in the 1960s and 1970s. The new instruments were expensive to purchase and maintain, so academic research became even more dependent on industrial and government financial support; interpretation of the data depended upon quantum mechanical considerations and this, coupled with the abandonment of traditional methods of wet and dry analyses, forced changes in the way that chemistry was taught. In France, for example, theories of atomic structure and the periodic table only made their appearance in secondary schools in 1978.

Teaching laboratories in schools and universities were reshaped in these decades, and syllabuses were reconstructed so that far greater emphasis was placed upon physical chemistry, electronic structure, and quantum chemistry. Synthetic chemistry was no longer a tedious process to confirm a possible structure, but became a logical way of strategically determining the structures of ever more complicated compounds, and even designing new materials from scratch. The understanding of reaction mechanisms was also revolutionized by the revelation of the roles played by 'unstable' intermediates during molecular transformations.

Finally, physical instruments opened up the possibility of better understanding of biochemical pathways in living systems and of revealing the extent of pollution in the environment. Regarding the latter, the delicacy of the chemists' instruments has made it possible for governments to legislate and monitor the dangerous products of our working worlds.

Chemistry's millennium

The British historian John Agar has interpreted the history of 20th-century science in terms of problems that arose in the everyday existence of human life such as warfare and health. Since the 1950s universities have ceased to be purely academic institutions ('ivory towers') devoted to the teaching and promotion of individual disciplines. In our contemporary 'post-academic' world they have become inter and transdisciplinary, as well as utilitarian, institutions. Like industry and business generally, universities now have to make money and are consequently directed by politics and economics.

This has been no bad thing for chemistry's fortunes insofar as it has made its central role in the circle of the sciences much more prominent; it has become an essential science. The Japanese historian of polymer science Yasu Furakawa has described molecular biology, for example, as a fusion of the methods, techniques, and concepts of organic chemistry, polymer chemistry, biochemistry, physical chemistry, X-ray crystallography, genetics, and bacteriology. Much the same might be said of the synthetic materials and technology that now dominate our computerized man-made world.

Epilogue

The introduction of electronic instruments into analytical and synthetic chemistry in the 1960s is an appropriate point to conclude this short account of the history of chemistry. Of course, the subject has not stood still during the past sixty years.

On the negative side, the chemists' wars of the 20th century had already coloured public perception of chemistry as a destructive science; this perception of chemistry's ambivalence was further exacerbated by a more general chemophobia produced by revelations of the pollution trails left by heavy industry, the widespread use of pesticides in agriculture and veterinary medicine, and the devastating explosion of a chemical plant in Bhopal in 1984.

While the historian can take a long-term view of this image problem, namely that chemistry has always been associated with risk and that explosions and pollution have been recurrent events, new analytical techniques and risk assessment legislation are undoubtedly having a remedial effect. Instruments like James Lovelock's miniature electron-capture detector of the 1950s played a major role in revealing the extent of man-made pollution and its alarming effects on the protective atmospheric ozone layer, and the potential of carbon dioxide emissions for disturbing climate.

More positively, computers replaced the laboratory notebook, and with their aid chemists began to study perfect three-dimensional structural models and design new molecules on the screen. This ability has led to new ways to craft drugs that can inhibit undesirable metabolic processes. Combinatorial chemistry was developed by pharmaceutical companies in the 1990s. By reacting together multiple ingredients rather than single ones, huge numbers of compounds and their derivatives were created in one reaction and the resultant 'library' of compounds screened for possible use as drugs. Pharmaceutical companies have spent millions of pounds on automating these syntheses and the preparation of individual compounds are only scaled up when a promising molecule is detected.

The American chemist L. R. Corey also introduced what he called retrosynthesis in the 1990s. In this technique, the substance whose structure is to be determined and synthesized is broken down into simpler molecules whose structures are known and whose individual synthesis has been already achieved. This knowledge can then be used to plan a synthesis of the wanted molecule. While mainly used in pharmaceutical research and drug design, these latest synthetic techniques have also had aesthetic appeal with syntheses being attempted and achieved purely for intellectual enjoyment.

Chemists have also continued to make extraordinary, surprising, and useful discoveries. Carbon, for example, was shown in 1985 to have an allotropic form different from its known varieties of graphite, diamond, and charcoal. This new form, which contained sixty atoms of carbon formed in the shape of a football, was cumbersomely named Buckminsterfullerene in honour of the American architect who had devised the similarly shaped geodesic dome. Fullerene and its derivatives (some in the shape of cylinders called nanotubes) quickly became the subject of much synthetic research.

Even more sensationally, in 2003 a one-atom-thick two-dimensional form of graphite was prepared consisting of flat sheets of hexagonally ringed carbon with valuable electrical conduction properties. Named graphene, it has been quickly put to work in the creation of new extremely light construction materials and faster semiconductors. By analogy, chemists have found that graphene-type allotropes can also be formed by phosphorus (phosphorene), silicon (silicone), germanium (germanene), and arsenic (arsenene). Needless to say, like isotopes a century ago, the existence, or creation, of such diverse allotropes raises the philosophical question as to what chemists mean by an element.

A new technology has emerged, nanotechnology, which refers to the manipulative transformation of matter at an atomic or molecular level. This was something that chemists had always done before at a macroscopic scale through analysis and synthesis; now this can be carried out at the nanoscopic scale, atom for atom. Here quantum effects become important and confer extraordinarily valuable properties on materials—optical, electrical, and structural. The resultant 'materials science' concentrates on the synthesis and construction of molecules of commercial, industrial, and pharmaceutical significance.

The exploration of this atomic material world involves not just chemists, but biochemists, physicists, and engineers. In an age of cross-disciplinary, transdisciplinary science and technology, several historians, sociologists, and philosophers of science have queried whether the concept of distinct scientific disciplines like chemistry, physics, and biology serves a purpose any longer. According to this view, chemistry has metamorphosed into biochemistry, biotechnology, nanotechnology, and materials science. Driven by technology, it has dissolved into a service science.

While there are sound arguments for this view, which has implications for the way the sciences are presently taught in schools and universities, practising chemists prefer to see chemistry as the central science that underpins the physical and biological sciences. Far from dissolving, it has, in effect, taken over these other disciplines. As Liebig said in his popular writings in the 1840s, 'Alles ist Chemie'—everything is chemistry.

Sources of quotations

Introduction

Friedrich Kekulé, *Lehrbuch der organischen Chemie* (Erlangen: F. Enke, 1861), p. 1.

Robert Siegfried, *From Elements to Atoms. A History of Chemical Composition* (Philadelphia: American Philosophical Society, 2001), pp. 1–2.

Chapter 1: On the nature of stuff

Robert P. Multhauf, *The Origins of Chemistry* (London: Oldbourne, 1966), p. 27.

The Journals of John Tyndall (typescript), Royal Institution of Great Britain, 6 February 1854.

Chapter 2: The analysis of stuff

William R. Newman and Lawrence M. Principe, *Alchemy Tried in the Fire* (Chicago: Chicago University Press, 2002), p. 180.

Chapter 3: Gases and atoms

Antoine Lavoisier, Memorandum of 20 February 1773, quoted in Henry Guerlac, *Lavoisier: The Crucial Year* (Ithaca, NY: Cornell University Press, 1961), p. 229.

Antoine Lavoisier, 'Mémoire sur la nature du principe qui se combine avec les métaux pendant leur calcinations', *Oeuvres de Lavoisier*, 6 vols (Paris: Imprimerie Impériale, 1862–93), vol. 2, p. 123.

Henry Cavendish, 'Experiments on air', *Philosophical Transactions of Royal Society*, 74 (1784), 119–53, at p. 129.

Sacha Tomic, *Comment la chimie a transformé le monde* (Paris, 2013), p. 64 (my translation).

Chapter 4: Types and hexagons

August Laurent, *Chemical Method* (London: Cavendish Society, 1855), p. 203.

Alexander Williamson, 'Theory of etherification', *Quarterly Journal Chemical Society*, 4 (1852), 229–39, at p. 239.

Edward Frankland, *Sketches from the Life of Edward Frankland* (London, 1902), p. 187.

Friedrich Kekulé, 'Ueber die Constitution des Mesitylens', *Zeitschrift für Chemie*, 10 (1867), 214–18, reprinted in Richard Anschütz, *August Kekulé*, 2 vols (Berlin, 1929), vol. 2, pp. 5–30.

Chapter 5: Reactivity

Staffan Bergwik, 'An assemblage of science and home: the gendered lifestyle of Svante Arrhenius', *Isis*, 105 (2014), 265–91, at p. 281.

Chapter 6: Synthesis

Justus Liebig and Friedrich Wöhler, 'Untersuchungen über die Natur der Harnsäure', *Annalen der Pharmacie*, 26 (1838), 241–340, at pp. 314–15.

Further reading

The principal reference work in English is James R. Partington, *A History of Chemistry*, 4 vols (Macmillan: London, 1965–70); but see also the more digestible work of Aaron J. Ihde, *The Development of Modern Chemistry* (New York: Harper & Row, 1964). More recent analytical accounts include William H. Brock, *The Fontana History of Chemistry* (London: HarperCollins, 1992; published in America as *The Norton History of Chemistry* (New York: Norton, 1993); and reissued as *The Chemical Tree* (New York: Norton, 2000); John Hudson, *The History of Chemistry* (Basingstoke: Macmillan, 1992) is aimed at practising chemists; David M. Knight, *Ideas in Chemistry: A History of the Science* (London: Athlone Press, 1992); Bernadette Bensaude-Vincent and Isabelle Stengers, *A History of Chemistry* (Cambridge, MA: Harvard University Press, 1996) previously published in French (Paris, 1993); Trevor H. Levere, *Transforming Matter: A History of Chemistry from Alchemy to the Buckyball* (Baltimore, MD: Johns Hopkins University Press, 2001); and Peter J. T. Morris, *The Matter Factory: A History of the Chemistry Laboratory* (London: Reaktion Books, 2015). A valuable source on chemical technologies (a subject that I have largely ignored) is Charles Singer, Eric J. Holmyard, and A. Rupert Hall, eds, *History of Technology*, 8 vols (Oxford: Clarendon Press, 1954–84). There are two international journals of the subject: *Ambix: The Journal of the Society for the History of Alchemy and Chemistry* (1937 onwards) and *The Bulletin of the History of Chemistry* published by the Division of the History of Chemistry of the American Chemical Society (1988 onwards).

Chapter 1: On the nature of stuff

James R. Partington, *Origins and Development of Applied Chemistry* (Longmans, 1935) is for reference only; more readable are Robert P. Multhauf, *The Origins of Chemistry* (London: Oldbourne, 1966) and Ronald F. Tylecote, *History of Metallurgy*, 2nd ed. (London: Institute of Materials, 1992); William Eamon, *Science and the Secrets of Nature: Books of Secrets in Medieval and Early Modern Culture* (Princeton, NJ: Princeton University Press, 1994); Philip Ball, *Bright Earth: Art and the Invention of Colour* (Harmondsworth: Penguin Books, 2001; Chicago: University of Chicago Press, 2003); William R. Newman, *Promethian Ambitions: Alchemy and the Quest to Perfect Nature* (Chicago: University of Chicago Press, 2004) and his *Atoms and Alchemy* (Chicago: University of Chicago Press, 2006); Joseph B. Lambert, 'The deep history of chemistry', *Bulletin History of Chemistry*, 30, no. 1 (2005), 1–9; Marco Beretta, *The Alchemy of Glass: Counterfeit, Imitation, and Transmutation* (Sagamore Beach, MA: Science History Publishing, 2009); and L. M. Principe, *The Secrets of Alchemy* (Chicago: University of Chicago Press, 2013). For an entertaining book that provides instructions and interpretations of alchemical experiments, see Cathy Cobb, Monty L. Fetterolf, and Harold Goldwhite, *The Chemistry of Alchemy* (Amherst, NY: Prometheus Books, 2014).

Chapter 2: The analysis of stuff

Allen G. Debus, *The English Paracelsians* (London: Oldbourne, 1965); Robert P. Multhauf, *The Origins of Chemistry* (London: Oldbourne, 1966); Frederic L. Holmes, *Eighteenth-Century Chemistry as an Investigative Enterprise* (Berkeley, CA: University of California Press, 1989); Antonio Clericuzio, *Elements, Principles and Corpuscles: A Study of Atomism and Chemistry in the Seventeenth Century* (Dordrecht: Kluwer, 2000); Philip Ball, *The Devil's Doctor: Paracelsus and the World of Renaissance Magic and Science* (London: Heinemann, 2006); William R. Newman, *Atoms and Alchemy: Chymistry and the Experimental Origins of the Scientific Revolution* (Chicago: Chicago University Press, 2006); Charles Webster, *Paracelsus: Medicine, Magic and Mission at the End of Time* (New Haven, CT: Yale University Press, 2008); Victor D. Boantza, *Matter and Method in the Long Chemical Revolution* (Farnham: Ashgate, 2013).

Chapter 3: Gases and atoms

Archie and Nan Clow, *The Chemical Revolution* (1952); Frank Greenaway, *John Dalton and the Atom* (London: Heinemann, 1966); Donald S. L. Cardwell (ed.), *John Dalton and the Progress of Science* (Manchester: Manchester University Press, 1968); Elisabeth Patterson, *John Dalton and the Atomic Theory* (New York: Doubleday, 1970); Arnold Thackray, *John Dalton: Critical Assessments of his Life and Science* (Cambridge, MA: Harvard University Press, 1972); William H. Brock, *Fontana History of Chemistry* (1992), chapter 4; Arthur Donovan, *Antoine Lavoisier. Science, Administration and Revolution* (Cambridge: Cambridge University Press, 1993); Mario Beretta, *The Enlightenment of Matter* (1993) for the development of chemical nomenclature; Maurice P. Crosland, *In the Shadow of Lavoisier: The Annales de Chimie and the Establishment of a New Science* (Faringdon: British Society for the History of Science, 1994); Robert E. Schofield, *The Enlightenment of Joseph Priestley* (University Park, PA: Pennsylvania State University Press, 1997) and its sequel *The Enlightened Joseph Priestley* (University Park, PA: Pennsylvania State University Press, 2004); Robert Siegfried, *From Elements to Atoms: A History of Chemical Composition* (Philadelphia, PA: American Philosophical Society, 2002); Jonathan Simon, *Chemistry, Pharmacy and Revolution in France, 1777–1809* (Aldershot: Ashgate, 2005); Victor D. Boantza, *Matter and Method in the Long Chemical Revolution* (Farnham: Ashgate, 2013).

Chapter 4: Types and hexagons

O. Theodor Benfey, *From Vital Force to Structural Formulas* (Boston: Houghton Mifflin, 1964); Colin A. Russell, *The History of Valency* (Leicester: University of Leicester Press, 1971); Alan J. Rocke, *The Quiet Revolution: Hermann Kolbe and the Science of Organic Chemistry* (Berkeley, CA: University of California Press, 1993); Colin A. Russell, *Edward Frankland: Chemistry, Controversy and Conspiracy in Victorian England* (Cambridge: Cambridge University Press, 1996); William H. Brock, *Justus von Liebig: The Chemical Gatekeeper* (Cambridge: Cambridge University Press, 1997); Alan J. Rocke, *Nationalizing Science: Adolphe Wurtz and the Battle for French Chemistry* (Cambridge, MA: MIT Press, 2001); and for those who read French, Sacha Tomic, *La pratique de l'analyse chimique et l'émergence de la chimie organique: une enterprise pluridiscipline (1790-1835)* (Rennes: Le Square, 2013).

Chapter 5: Reactivity

M. M. Patterson Muir, *A History of Chemical Theories and Laws* (New York: J. Wiley, 1907; reprint New York: Arno Press, 1975); J. W. van Spronsen, *The Periodic System of Chemical Elements* (Amsterdam: Elsevier, 1969); C. A. Russell et al., *Chemistry as a Profession* (Milton Keynes: Open University, 1977); John Servos, *Physical Chemistry from Ostwald to Pauling: The Making of a Science in America* (Princeton, NJ: Princeton University Press, 1990); William H. Brock, *The Fontana History of Chemistry* (London: HarperCollins, 1992), chapter 10; Mary Jo Nye, *From Chemical Philosophy to Theoretical Chemistry* (Berkeley, CA: University of California Press, 1993); Keith Laidler, *The World of Physical Chemistry* (Oxford: Oxford University Press, 1995); Eric Scerrie, *The Periodic Table: Its Story and its Significance* (Oxford: Oxford University Press, 2007); Peter Atkins, *Physical Chemistry: A Very Short Introduction* (Oxford: Oxford University Press, 2014).

Chapter 6: Synthesis

Lutz F. Haber, *The Chemical Industry 1900-1930* (Oxford: Clarendon Press, 1971); Anthony S. Travis, *The Rainbow Makers: The Origins of the Synthetic Dyestuffs Industry in Western Europe* (Bethlehem, PA: Lehigh University Press, 1993); Roy MacLeod, 'Chemists go to war', *Annals of Science*, 50 (1993), 455–81; Roald Hoffmann, *The Same and Not the Same* (New York: Columbia University Press, 1995); Yasu Furukawa, *Inventing Polymer Science* (Philadelphia, PA: University of Pennsylvania Press, 1998); Peter Morris, ed., *From Classical to Modern Chemistry: The Instrumental Revolution* (Cambridge: Royal Society of Chemistry, 2002); Marelene and Geoff Rayner-Canham, *Chemistry was their Life: Pioneer British Women Chemists 1880-1949* (London: Imperial College Press, 2008); John Agar, *Science in the Twentieth Century and Beyond* (Cambridge: Cambridge University Press, 2012); Michael Sutton, 'Chemists at war', *Chemistry World*, 11 (August 2014), 44–7; Mark Miodownik, *Stuff Matters: The Strange Stories of the Marvellous Materials that Shape our Man-made World*

(London: Viking, 2014); Catherine Jackson, 'Synthetical experiments and alkaloid analogues: Liebig, Hofmann, and the origins of organic synthesis', *Historical Studies in Natural Science*, 44 (2014), 319–63. Although it demands some technical knowledge, a readable history of quantum chemistry is Kostas Gavroglu and Ana Simões, *Neither Physics nor Chemistry* (Cambridge, MA: MIT Press, 2012).

Index

Index

SOCIAL MEDIA
Very Short Introduction

Join our community

www.oup.com/vsi

- Join us online at the official Very Short Introductions
  Facebook page.
- Access the thoughts and musings of our authors with our
  online **blog**.
- Sign up for our monthly **e-newsletter** to receive information
  on all new titles publishing that month.
- Browse the full range of Very Short Introductions online.
- Read **extracts** from the Introductions for free.
- Visit our library of **Reading Guides**. These guides, written by our
  expert authors will help you to question again, why you think
  what you think.
- If you are a teacher or lecturer you can order inspection
  copies quickly and simply via our website.